Enzymes: A Very Short Introduction

Very Short Introductions available now:

Available soon:

For more information visit our website

www.oup.com/vsi/

Paul Engel

ENZYMES

A Very Short Introduction

OXFORD
UNIVERSITY PRESS

OXFORD
UNIVERSITY PRESS

Great Clarendon Street, Oxford, OX2 6DP,
United Kingdom

Oxford University Press is a department of the University of Oxford.
It furthers the University's objective of excellence in research, scholarship,
and education by publishing worldwide. Oxford is a registered trade mark of
Oxford University Press in the UK and in certain other countries

Published in the United States of America by Oxford University Press
198 Madison Avenue, New York, NY 10016, United States of America

British Library Cataloguing in Publication Data
Data available

Library of Congress Control Number: 2020941867

ISBN 978-0-19-882498-5

Printed in Great Britain by
Ashford Colour Press Ltd, Gosport, Hampshire

Contents

List of illustrations

Enzymes

Acknowledgements

I must thank several people who encouraged and helped me to write this book. First my wife Sue for setting me on the path in the first place. Next my sons Tom and Ben, who gave valuable criticism of style and content, and Dr Margaret Deith who cast an expert editor's eye over the early chapters. I was both grateful and relieved when my colleague Professor Des Higgins gave Chapter 6 a clean bill of health after correcting some details. Finally I must thank Dr Latha Menon and Ms Jenny Nugee at OUP along with Joy Mellor, Govindasamy Bhavani, and Dorothy McCarthy who kept me on the straight and narrow and nevertheless, by their speed, efficiency and good humour, made it a happy publishing experience.

Chapter 1
No enzymes, no life

What is life? We take it for granted most of the time. We are part of it, we are surrounded by it. The dog tearing across the park, the caterpillars attacking your cauliflower plants, the cauliflower plants themselves, it usually seems fairly obvious whether things are alive or not. But if we stop and try to put into words what distinguishes living things from the rest of a dead world, it becomes surprisingly complicated. We can try to define life functionally, that is in terms of what living things do, highlighting their ability to grow, to gather in the raw materials that make it possible to grow, to reproduce, to respond and adapt to their environment, etc. We can also think about living things in terms of their structure, noting the extraordinary variety between different forms of life and extraordinary complexity within each one. If we then use the tools of the chemist, we find that living organisms are made of extremely complex building blocks. Nevertheless, there is an underlying and striking similarity between the complex substances that make up different forms of life, say a spider and a cabbage. We can also take apart living cells to see if we can find the fundamental and distinctive working parts that make life processes possible. In this latter regard, most people are aware of the importance of DNA (deoxyribonucleic acid), to the extent that it has entered everyday vocabulary (usually in entirely inappropriate contexts). DNA carries the instructions for life. These can be implemented within an organism and can also be duplicated and passed on from generation to generation of organisms or cells.

Crucial though it is, DNA on its own is not life, however. Left to its own devices, it would do nothing and go nowhere. A cookbook recipe does not set the meal on the table. In the same way, to carry out the instructions encoded in the DNA and then to drive all the varied things that we recognize as life, some kind of machinery is needed to make it all happen. This active, executive role belongs to *enzymes*. But what are enzymes and, beyond the vague statement just made, what do they do?

It is clear, first of all, that when enzymes go wrong it is a serious matter. Many distressing genetic diseases, handed on from parent to child, are due to defective enzymes. Because of their crucial role, enzymes are also often targets for deadly poisons used as insecticides and herbicides but also sometimes as chemical weapons. In March 2018, a former Russian spy and his daughter were mysteriously poisoned on the streets of Salisbury, England, by a chemical smeared on the doorknob of their house. News reports explained that the Novichok agent acts by attacking an enzyme. A few weeks later, further reports attributed the pair's miraculous survival to the fact that, by supporting vital functions such as breathing and heart contraction, the clinicians had provided time for new enzymes to be made.

Most people have probably met enzymes in school biology as the agents responsible for digesting our food, breaking down the starch of pasta, rice, potatoes into sugar and so on. Many meet them again as they face their washing machine, stained sports clothes in hand, and wonder whether or not to use a 'biological' detergent, containing added but unspecified 'enzymes' to do mysterious things to the clothes. As it happens, in both contexts the enzymes' function is very similar, breaking down large chemical molecules into smaller bits that will wash away. People do not generally realize, however, that enzymes have much wider and more diverse roles and that, in effect, they orchestrate the whole of life.

Almost every chemical step in every living thing is guided by its own dedicated enzyme. Out of all the multitude of reactions that might theoretically be possible, enzymes select and guide the orderly sequences of reactions that we call 'metabolism', breaking down foodstuffs stepwise to give useful building blocks and reassembling them to make new biological molecules. Whether the sequence of reactions turns carbon dioxide, water, and mineral salts from the soil into a tree or brings about the emergence of a walking, talking human adult from a single fertilized cell, one after another the jigsaw pieces fall into place, unerringly steered by the team of enzymes. Enzymes harness the locked-up energy in some of our foodstuffs or, in the case of plants, the energy of sunlight, and redeploy that energy to drive a vast range of biological tasks. They switch processes on and off. They even make and interpret the DNA instructions. Finally, what makes enzymes themselves? The answer, of course, is enzymes.

Nowadays, because enzymes literally control every biological process, many of our drugs are designed to knock out or damp down a particular enzyme activity. They are also increasingly used on their own, outside of their original biological context, as tools, for example in diagnostic kits, in chemical synthesis, in processing foodstuffs, leather, woodpulp, etc. We shall return to some of these important applications in later chapters.

This, however, raises many questions. If we are going to talk about life in terms of sequences of chemical reactions, we need to ask whether living things carry out the kind of 'ordinary' chemistry that we would recognize from our school science lessons. Can 'ordinary' chemistry produce the seemingly special substances and structures characteristic of life? And at an even more basic level, what determines whether a chemical reaction can or does occur? How can enzymes influence these processes? How do they all fit together to make up something we can recognize as life?

Beginnings and basics

The smallest unit of a substance is the molecule. Molecules may be combinations either of a single kind of atom or of two or more different kinds of atom. In the molecules of a pure substance, atoms are joined in precise numerical combination. In oxygen, for example, two oxygen atoms, each given the chemical symbol O, join together to make an oxygen molecule, represented as O_2. In black copper oxide (CuO), one atom of shiny copper (Cu) is combined with one atom of oxygen (O). Similarly, in every molecule of table salt, one atom of the metal sodium (Na) is combined with one atom of the gas chlorine (Cl), making sodium chloride (NaCl), which resembles neither sodium nor chlorine—fortunately, since both sodium and chlorine on their own would be decidedly dangerous substances to sprinkle on your food. The formula NaCl tells you the combining proportions, one to one in this case. In water molecules, the proportions are different, since a single atom of the gas oxygen combines with two atoms (H) of the gas hydrogen to form water (H_2O). This unequal recipe reflects the fact that oxygen has a higher combining power than hydrogen, a property that chemists refer to as *valency*. Hydrogen, chlorine, and sodium all have valencies of one, whereas oxygen and copper have valencies of two. These different combining powers relate to the underlying structure of the individual atoms, a level of complexity we need not delve into, and the simplest way to think about valency is as the number of 'arms' each atom has to hold hands with other atoms. Oxygen (O) has two arms but hydrogen (H) has only one.

In a chemical reaction the atoms in the molecules involved rearrange into new combinations. Thus, in the specific example just mentioned, each atom from a molecule of oxygen (O_2) can combine with two atoms of hydrogen. Therefore, for the arithmetic to work out correctly, the two O atoms in a single molecule of oxygen (O_2) interact with the total of four H atoms in *two* molecules of hydrogen (H_2) forming two new molecules of water.

$$2 H_2 + O_2 \rightarrow 2 H_2O$$

The equation for this reaction reveals that all the atoms on the left-hand side (four hydrogen atoms and two oxygen atoms) reappear, though differently arranged, on the right-hand side. Normally all the atoms of the molecules involved in a chemical reaction are conserved.

Discovery of the rules that govern the combination and rearrangement of atoms led to rapid progress through the 19th century and to the development of a large chemical industry to make useful new substances—disinfectants, solvents, anaesthetics, dyes, explosives, etc. Carbon (C), in particular, turned out to be the basis for making a very large number of different types of molecule because, unlike H or O, the carbon atom has a valency of 4. Its ability to 'hold hands' with several atoms at once allows it to form complex chains and rings and networks of linked atoms (Figure 1). As we shall see later, this property of carbon turns out also to be fundamental to the chemistry of life.

Can chemistry explain biology?

For a long time many chemists were convinced that living things do not fit into the rational framework of conventional chemistry. Fundamentally, there are two issues: are the substances found inside living organisms recognizable 'proper' chemicals; and, second, if so, are the processes of life recognizable chemical processes? It was clear that living things could take in real chemicals and convert them ultimately to real chemical products. For example, with the aid of oxygen, animals convert glucose

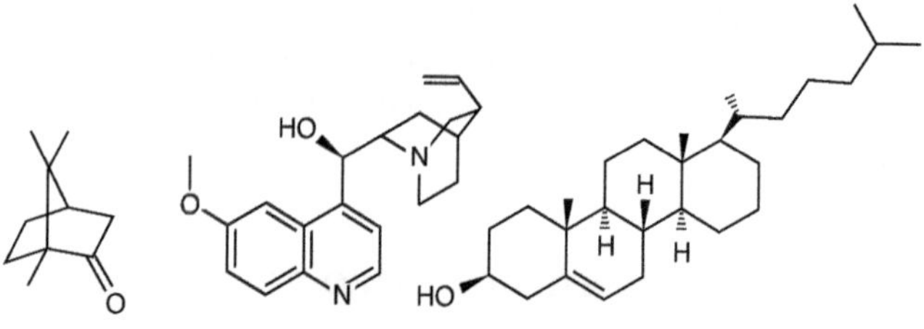

Camphor   Quinine   Cholesterol

Complex bio-organic molecules

1. **Representative organic molecules. The three shown, all biologically
produced, indicate the complexity of structure made possible by
carbon's valency of 4. By convention, corners, ends, and crossings
where no atom is shown are carbon atoms, and, where a C appears to
have fewer than four bonds, the missing ones are to hydrogen atoms
not shown.**

eventually to carbon dioxide and water. We now know that this
does not occur in a single chemical reaction, but is the cumulative
outcome of a series of reactions. Nevertheless, it can be
represented by a respectable chemical equation for the overall
conversion:

$$C_6H_{12}O_6 \ + \ 6\,O_2 \rightarrow \ 6\,CO_2 \ + \ 6\,H_2O$$

glucose   oxygen carbon dioxide water

All the input and output substances are well-recognized
chemicals. In the same way, in the absence of oxygen, yeast
converts the same starting sugar, glucose, to another well-known
chemical, ethyl alcohol (ethanol), C_2H_5OH.

$$C_6H_{12}O_6 \rightarrow \ 2\,C_2H_5OH + 2\,CO_2$$

glucose   ethanol   carbon dioxide

Nevertheless, in the 1820s the eminent English chemist William
Henry confidently asserted that

6

It is not probable that we shall ever attain the power of imitating Nature in these operations. For, in the functions of a living plant, a directing principle appears to be concerned, peculiar to animated bodies, and superior to and different from the cause that has been termed chemical affinity.

Only a few years later, in 1828, the German chemist Friedrich Wöhler experimented with ammonium cyanate, a known simple 'inorganic' 'chemical' compound. When it was heated, it formed crystals which he recognized as being *urea*. This is a clearly 'biological' compound, which can be extracted from urine but in this case was made, as he reported, 'without the help of a man, a dog or a kidney'. The molecules of the two substances have exactly the same eight atoms, just in different arrangements. There was no hint, therefore, of a different kind of matter, obeying different rules.

On the other hand, accepting that biological substances were probably just part of the overall pattern of chemistry, and that biological organisms carry out chemical conversions, still gave no clue as to exactly *how* living things carry out those conversions. How glucose was oxidized or fermented, how urea was produced, how indeed any biological conversion was achieved remained a complete mystery until the 20th century. Also, despite the acceptability of glucose, ethanol, carbon dioxide, urea, etc., as authentic chemicals, there still seemed to be much within living cells that could hardly expect diplomatic recognition among chemists. Chemists expected well-behaved substances that would cleanly obey the laws of mathematics and that, like the compounds they made in their laboratories, would form crystals, display precise melting and boiling points, etc. As well as 'proper' chemicals like urea and ethanol, biology appeared to be also populated by various unruly, strange, and sticky, complex substances with special properties and different rules (Box 1). Was this, after all, a totally different realm, chemically speaking?

Box 1 Organic chemistry

The 19th-century confusion over the relation between biology and chemistry is reflected in the varied use to this day of the word 'organic'. Chemists sought to analyse and identify the complex carbon compounds produced by living things and then to make them in the laboratory. Since the chemicals were derived initially from living organisms, this logically became 'organic chemistry'. This term, however, rapidly expanded to embrace the whole of carbon chemistry including complex non-biological molecules. In 1856, for example, William Perkin's attempts to make the antimalarial plant compound quinine synthetically in the laboratory led instead, by accident, to the production of a dye, Perkin's mauve, laying the foundations of a highly profitable chemical dye industry in both Britain and Germany. All scientists gratefully accept lucky accidents, but systematic modern organic chemistry does not only rely on accident and has seen over a century of success leading to the production of synthetic drugs, detergents, pesticides, plastics, textiles, anaesthetics, paints, adhesives, etc. This has led ultimately to the contemporary confusion in which 'organic' foods are those in which 'organic' pesticides, hormones, drugs, etc., have *not* been used in their production.

The main controversy, however, through much of the 19th century, centred not on the substances but on the processes, and in particular those involved in the production of beer and wine. In both cases sugar is fermented to produce alcohol. The sugar for wine-making is present in the grapes whereas, in brewing, the sugar is made in a 'malting' process. This converts the starch of the initial grain feedstock to sugar (glucose). The eminent German chemist Justus von Liebig maintained that the interconversion of these recognizable chemicals must be a chemical process and that there must be chemical agents within the cell to bring it about. His great adversary was the French biologist Louis Pasteur, whose experiments played a big part in establishing the importance of living cells in these fermentation processes. Pasteur also showed that living cells were responsible for 'spoilage' of wine, juices, milk, etc., and that such spoilage can be prevented by gentle heating, which kills the organisms responsible. This process, 'pasteurization', is still used in the food industry today, especially for dairy products and fruit juices. Pasteur believed he had established a fundamental truth, namely that for biological processes to occur there must be living organisms. He thus became one of the champions of '*vitalism*', a belief that biological processes depended on a mysterious 'vital force' and were therefore inseparable from life.

In fact, as early as 1833, Payen and Persoz had shown that they could treat malt extract with alcohol and precipitate a substance that they called 'diastase' which had the ability to convert starch to sugar. This was clearly in conflict with the vitalist idea, because diastase is a non-living agent. It is prepared from previously living material but it no longer contains anything alive. However, as often happens in science, this was a discovery whose significance was not fully appreciated at the time and lay unused until the end of the century.

The death of vitalism and the birth of biochemistry

Despite the early work of Payen and Persoz, vitalism was not finally laid to rest until 1898, when the Bavarian brothers, Eduard and Hans Büchner, announced that not only living yeast cells but also the cell-free juice extracted from broken yeast cells could carry out fermentation. Thus, undeniably, a complex biological process could take place without any living organisms being present. No need then for a mysterious vital force. The resolution of this key issue finally set the stage for the emergence of a formal discipline of *biochemistry* with its own learned journals and degree courses. Over and over again, in the years that followed, non-living extracts from all sorts of living sources were shown to carry out complicated chemical conversions. These conversions frequently involved a long sequence of many separate chemical steps. When this is so, we call it a *metabolic pathway*. For each step there is a dedicated, specific agent in the extract that makes that individual reaction possible—an enzyme. It is the existence of precisely the right orchestra of enzymes that makes possible the ordered chemistry that we call life.

Each biological extract is thus like a complex soup containing a mixture of many different potent agents. Over the first decades of the 20th century various sorting procedures were deployed in order to separate these components. This often-arduous *'fractionation'* process ultimately provided the different individual enzymes in pure forms. This meant they could be studied separately, without confusion or interference from other enzymes, to find out unambiguously what each one did. These studies led, step by step, towards an understanding of how living things work.

Despite this steady progress, biochemistry had to struggle for many years to achieve academic recognition. Nineteenth-century prejudices persisted, and chemists, secure in the accomplishments of their own discipline, resisted the pretensions of the upstart biochemistry newcomer. So, for example, in the 1940s, Hans Krebs, one of the scientists who had fled Germany before World War II, proposed a degree course in biochemistry to the Science Faculty at the University of Sheffield. Reputedly, in response, his eminent opposite number in Chemistry asked, 'Gentlemen, why would we want a degree in biochemistry when everyone knows that biochemistry is merely chemistry done badly?' At this stage, Krebs had already discovered (at Sheffield) the central metabolic cycle that bears his name and is today familiar to every student of biology and won him the Nobel Prize in 1953. The interdisciplinary acrimony was still destructively evident when this author joined the Biochemistry Department at Sheffield in 1970.

Despite such prejudice, the application of physics, chemistry, and open minds to biological systems over the past hundred years has brought a staggering advance in our knowledge and understanding of almost every life process. A key element of this understanding is the realization that for each biological task there is a unique, exquisitely tailored set of enzymes. Already in 1932, another biochemistry Nobel Laureate, Sir Frederick Gowland Hopkins, was moved to state, in his presidential address to the Royal Society, 'It is, I think, difficult to exaggerate the importance to biology, and, I venture to say, to chemistry no less, of extended studies of enzymes and their action.' Extended studies have gone on apace ever since and, as I hope to show in later chapters, from today's vantage point Hopkins' pronouncement seems a cautious understatement.

Chapter 2
Making things happen—catalysis

Enzymes' role is to *catalyse* particular chemical reactions—that is, to speed them up. This process is not unique to living systems (or extracts prepared from living systems), although enzymes *are* both more potent and more selective than catalysts encountered elsewhere in chemistry. Chemical catalysis was first observed and commented upon in 1794 by the Scotswoman Elizabeth Fulhame. She noticed that, for a number of reactions with oxygen, such as the conversion of carbon monoxide (CO) to carbon dioxide (CO_2), water needed to be present. Water was somehow involved in the reaction but not consumed. All the water was still there when the reaction was complete. This description gives us what is still the accepted essential definition of a catalyst as *an agent that speeds up a chemical reaction but remains unchanged itself at the end of the process*. A little later, in 1812, a German chemist, Gottlieb Kirchhoff, observed that chemical conversion of starch to sugar only proceeded in the presence of acid but that the acid remained unaltered. In 1817, Sir Humphry Davy, studying the reaction of flammable gases with oxygen, found that, in the presence of platinum, no flame was needed and reaction would proceed efficiently at temperatures as low as 50°C.

In the three examples above, a substance apparently unrelated to the reaction nevertheless effects a dramatic acceleration. In 1835, the Swedish chemist Jöns Berzelius, recognizing the common thread, coined the term *catalysis*, attributing it to 'an internal force whose nature is unknown to us'.

Since a catalyst is not altered or used up, it can be used over and over again. Thus, 200 years on, even though platinum is a costly precious metal, platinum catalysis is still widely used. Probably its most familiar application is in the 'catalytic converter' which efficiently turns unburnt hydrocarbons in motor exhaust (plus oxygen) into carbon dioxide and water vapour. More generally, once the concept of catalysis was established, examples and applications multiplied, leading to large-scale industrial exploitation as the 19th century advanced.

To the casual observer, a catalyst seems to make the difference between a reaction not happening at all without the catalyst and proceeding rapidly in its presence. However, strictly speaking, the difference is likely to be between a reaction too slow to notice and one that is rapid. The catalyst might, say, make the difference between a reaction requiring a minute to complete rather than a year. Usually we cannot be bothered to watch an uncatalysed reaction dawdling over a year, but measurements by exceptionally patient scientists allow us to determine the *rate enhancement*, revealing whether the reaction goes, say, a thousandfold faster with the catalyst or maybe a millionfold or more.

Thermodynamics and kinetics

We need now to tease this apart further and ask exactly what is happening. If a reaction is possible why doesn't it just 'go'? In truth, there are two distinct questions to disentangle:

1. What determines whether a chemical reaction is actually possible?
2. If it is possible, what determines how fast it will go?

Given these two basic questions, where does catalysis fit in? Can a catalyst make possible a reaction that is otherwise impossible?

First of all, we must rule out any notional reactions that are just chemical or mathematical nonsense, with different numbers of atom or kinds of atom on the two sides of the equation (we no longer believe in alchemy, and chemists need to be good accountants—atoms are not allowed to suddenly appear or disappear), or conversions that envisage unlikely rearrangements of multiple chemical linkages. We start by assuming that chemically our reaction makes good sense. Even so, this is no guarantee that it will proceed, because, like a ball on a slope, a chemical reaction can only go in the direction that involves a *loss of energy*. A ball will not spontaneously roll up a slope, and the same principle applies to chemical reactions. Whether the hypothetical reaction involves a loss of energy or not depends on the inherent properties of each of the reacting substances and potential products, the concentrations of each of these substances, and also the temperature. These considerations relating to the movement of energy are the realm of what physical chemists call *thermodynamics*.

However, even if a reaction is energetically feasible, it still may not go at a perceptible rate. Without the platinum, Davy's flammable gases were not going to react measurably with oxygen at 50°C. Even dynamite requires moderate persuasion to explode. Pursuing the ball analogy, we may imagine a small hollow in the slope where the ball lodges until a human foot or a puff of wind nudges it out to roll swiftly downhill (Figure 2). In chemical systems similarly, we talk of *energy wells*, in which potentially reactive molecules remain unchanged, and *transition states* from which they are able to react (corresponding to the lip of the small hollow, where the ball may either fall back into the hollow or roll down the slope). For reaction to occur, molecules need to be kicked into the transition state, allowing them then to run downhill by completing the chemical reaction. The 'kick' is

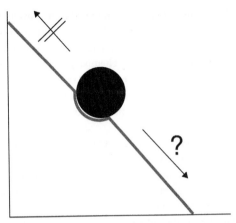

2. **Ball on slope. The ball will never spontaneously roll uphill, but even downhill it will only move when nudged out of its hollow.**

referred to as the *activation energy*. These considerations, governing how fast the reaction will go in reality, are the realm of *kinetics*.

Activation energy

To explain this more fully, let us imagine two substances, A and B, potentially able to react with one another and dissolved in the same vessel of water. There will be molecules of A and B distributed at random throughout the solution. Although the solution may look entirely still, at a submicroscopic level these molecules are all moving in random directions and at a variety of speeds. If substance A is going to react with substance B, their molecules have to meet, and so molecules of A have to bump into molecules of B. How often this happens will depend on the concentration of both substances, but, even when the molecules collide, they may not react, either because they meet in the wrong orientation or because they are not travelling fast enough to bump into one another with sufficient energy. One way to improve the

15

chances of reaction is simply to heat the mixture. This makes all the molecules travel faster and so a higher proportion of collisions will occur with the necessary activation energy. This accounts for the rule-of-thumb expectation in chemistry that the rate of a reaction will roughly double for every 10°C increase in temperature. (This approximation works because activation energies for most reactions fall within a relatively narrow range of values.)

There are also simpler, 'unimolecular' reactions, in which no collisions are required because the reaction just involves one kind of molecule transforming into another kind of molecule. In this case one may envisage most of the molecules contented in their 'energy well'. Only a small percentage of the molecules at any given moment will have sufficient energy to reach the transition state, in which they may tip over into chemical reaction, but again that percentage can be increased by heat.

Catalysis works by lowering the activation energy without the need to apply any heat. It offers an alternative route to the destination, just as a tunnel reaches the other side of a hill without going over the top. As already mentioned, a crucial part of the definition is that (like a railway tunnel) a catalyst remains unchanged once the reaction has occurred. This leaves it free to carry out the favour for more molecules, so that one molecule of the catalyst may process thousands of reaction cycles over a short period of time and, in theory, do so indefinitely (assuming a limitless supply of molecules to react).

Enzymes are biological catalysts but differ from ordinary chemical catalysts in two respects. First, they are very *specific*. A chemical catalyst that accelerates reactions of, say, an alcohol such as methanol is likely to work similarly with a wide range of different alcohols—ethanol, propanol, butanol, etc. An enzyme might only work with one or two very similar alcohols. An organism that needs to handle methanol (with one carbon atom) and propanol (with three carbon atoms) similarly, would usually have two separate enzymes, each one fine-tuned for its particular task. Second, enzymes are dramatically more *potent* than conventional chemical catalysts. The enzyme *carbonic anhydrase*, for example, catalyses the interconversion of bicarbonate ions and carbon dioxide and is important in disposing of CO_2 in our lungs. It speeds up the spontaneous uncatalysed reaction approximately ten-millionfold. This compares with rate enhancement of perhaps ten-thousandfold for a very good chemical catalyst.

Reversibility and equilibrium

At this point we need to introduce another concept, namely, the *reversibility* of reactions. If a reaction is energetically feasible, one might assume that it will proceed until one of the reacting substances is entirely used up. This is not so. Our imaginary reaction between A and B will make products. Let us call these C and D. As the reaction proceeds, the concentrations of A and B decrease and those of C and D increase. Two things will progressively happen. The rate of A reacting with B will decrease, but also, as the concentrations of C and D increase, the molecules of C will collide with those of D more and more frequently, and some will react to give A and B so that the reaction is going in opposite directions simultaneously. As the forward reaction slows and the reverse reaction speeds up, eventually the two reactions exactly balance. At this point, there is no net conversion in either direction and the reaction is said to be at *equilibrium*. Returning to our rolling ball, this tells us that to have a proper analogy, rather than a uniform slope, as in Figure 2, a chemical reaction is better represented as a curve with a slope gradually declining until finally it is horizontal. The curve must also climb up again on the other side, representing the situations in which the reaction would go in the reverse direction (Figure 3).

The point at which the reaction is at equilibrium, balanced so there is no net conversion, is a fundamental physical property of the reacting substances, entirely independent of catalysis.

3. Two-way curved slope. On either side of the slope the ball will run downhill (in opposite directions) to the lowest point.

A catalyst, enzyme or not, cannot determine the direction of reaction or how far it will go, but only how *fast* it proceeds towards equilibrium. An enzyme or any other catalyst, therefore, has to be able to accelerate a reaction in *either* direction. The direction in which the reaction proceeds under any given set of conditions will depend on various considerations but most notably the concentrations of all the reactants. Many biological reactions will go in either direction depending on the physiological situation. A good example is the reaction that produces lactic acid in our muscles during vigorous exercise (from a compound called pyruvic acid). This accumulation is what causes the temporary pain, for example, at the gym. In muscles at rest, or in other tissues of the body even during exercise, the same enzyme-catalysed reaction will go in the opposite direction, converting lactic acid back to pyruvic acid.

Gaining insight from enzyme kinetics

How do enzymes work? The old vitalist ideas persisted even after the Büchners' findings with cell-free extracts. It was suggested that enzymes still carried with them some of the 'vital force' of the living cell and that perhaps they radiated some of this force, acting on their target molecules at a distance. The only way to test and confirm or refute such ideas was experimentation. In the first few years of the 20th century, there was still no information about the chemical nature of enzymes themselves, but this did not prevent careful systematic studies of their effect and these proved to be very revealing. To measure the speed of a reaction one needs an accurate way of measuring the concentration either of the reacting substance (or one of the substances if there is more than one) or alternatively the concentration of (one of) the product(s). In our notional reaction of A with B to form C and D, it does not matter whether we measure the decrease in concentration of A or B or the increase in concentration of C or D. These increases/decreases must all be equal. If we are lucky, one of the reactants may have a distinctive property that can be measured in real time, so that the reaction can be observed as it happens. Often, however, the only possibility is to take timed samples, The reaction must be immediately stopped in each sample, so that concentrations do not change any more, and then the samples are analysed to determine the concentration of one of the reactants at each time point. Measurements of this kind allow us to determine the *initial rate*, that is, how fast the concentrations are changing before the reaction starts to slow down as it moves towards equilibrium (Figure 4).

With this methodology in place one can now start to ask how the initial rate varies as conditions are changed. Two experiments of this kind are informative. First, one can add different amounts of

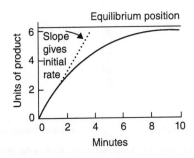

4. **Reaction time-course.**

the enzyme preparation to a series of identical reaction mixtures. This tells us that the rate is proportional to the amount of enzyme added; three times as much enzyme makes the reaction go three times as fast. What if we use a constant amount of enzyme but start the reaction with different concentrations of the reacting substance? With a simple chemical reaction, without any catalysis, if you double the concentration of a reacting substance you are obviously doubling the chance of molecules reacting in any chosen time interval, so that, in theory, if you could go on increasing the concentration, the reaction rate would increase indefinitely in exact proportion. This simple proportionality is indeed what is usually seen for an uncatalysed chemical reaction (Figure 5(a)).

Enzyme-catalysed reactions behave quite differently. This issue was studied by scientists in several countries in the two decades after 1890. In particular, attention was given to a reaction in which sucrose is split by water. Sucrose is just 'sugar', as in the sugar bowl on your table, but chemically it is classed as a so-called *disaccharide* because there are simpler, smaller sugar compounds (*monosaccharides*), and the sucrose molecule has two monosaccharides, glucose and fructose, linked together (Figure 6). If the linkage is attacked by water, this releases the monosaccharides. Without catalytic assistance this reaction would go exceedingly

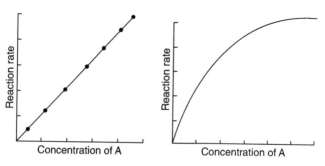

5. **Reaction rate vs reactant concentration: (a) first-order chemical reaction; (b) enzyme-catalysed reaction.**

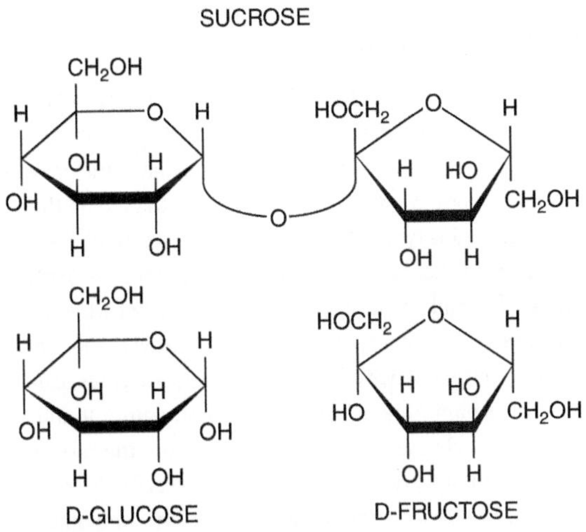

SUCROSE

6. Sugars—glucose, fructose, sucrose. The disaccharide sugar sucrose is formed by linking C1 of the monosaccharide glucose to C2 of fructose via an oxygen atom.

slowly, but it goes briskly in yeast cells or juices thanks to the action of an enzyme called *invertase* (Box 2).

The term *substrate* is used by enzymologists to denote whatever substance(s) an enzyme acts upon, so that, in our present example, sucrose is the substrate for invertase. Returning, then, to the kinetics, it was found that, with increases in the concentration of the substrate, the rate of the reaction with water did not go on and on upwards but reached a finite limiting rate (Figure 5(b)). In other words, beyond a certain sucrose concentration any further increase had no effect on the reaction rate. How could this be? Several scientists offered a possible explanation. Perhaps catalysis was brought about, not by long-distance vibrations but by a direct physical interaction between an enzyme molecule and a substrate molecule, forming what we refer to as an *enzyme-substrate (E-S) complex*. If catalysis only occurs in this complex, then the

21

Box 2 Naming enzymes

Early on in biochemistry, the convention was established of giving newly discovered enzymes names ending in 'ase', with the rest of the name telling you either what substance the enzyme acted upon or else what type of reaction it catalysed. In this case, chemists term this type of reaction a sugar 'inversion', hence invertase for the name of the enzyme. Except for a few early examples that had already slipped through the net with names such as trypsin or papain, all enzymes are named in this way. We should also note that, within this convention, enzymes usually have a 'trivial name' and a formal 'systematic name'. In this book we shall invariably use the trivial names because they are relatively short and are the ones all biochemists use in everyday discussion. The trivial names do also leave some ambiguity as to exactly what reaction is catalysed. This is why the formal names are needed, but they are not routinely used because they are so cumbersome. For example, the name invertase tells a biochemist roughly what biological job that enzyme does. The systematic name β-D-fructofuranoside fructohydrolase is a good deal more precise and informative but is rather a mouthful.

achievable rate will be limited by how many enzyme molecules are present. An analogy might be a city's taxi service. If there are 500 taxis but only 5 or 10 or 15 people looking for a taxi, then the number of journeys will be dependent on the number of people. On the other hand, if 5,000 or 10,000 or 15,000 people need a taxi, the queues may get longer and longer, but the number of journeys will be unchanging, as it is entirely limited by the number of taxis (Figure 7).

The most complete analysis, both experimentally and theoretically, of the dependence of the invertase reaction on

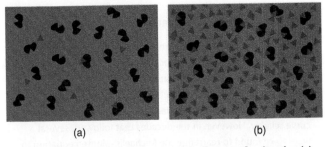

(a)　　　　　　　　　(b)

7. **Saturation of an enzyme. The two panels illustrate the situation (a) with substrate concentration a long way below K_m; and (b) approaching saturation (V_{max}).**

sucrose concentration was reported by Michaelis and Menten in 1913. Maud Menten was one of the first Canadian women to earn an MD degree, but, unable at that time to carry out research in Canada, she moved to Berlin to work with the eminent Professor Leonor Michaelis. Together they offered a simple mathematical treatment of the invertase reaction, leading to the *Michaelis–Menten equation*:

$$v = \frac{V_{max}[S]}{K_m + [S]}$$

In this equation the two variables are the substrate concentration $[S]$, which we set, and the reaction rate or velocity, v, which we measure (Box 3). What emerges from the mathematics is that for a particular set of experimental conditions we can describe an enzyme's behaviour in terms of two numbers. The first, V_{max}, is the limiting rate that is approached as you keep increasing the substrate concentration (as predicted by our taxi analogy). The second is the *Michaelis constant*, abbreviated as K_m. We could see it as a measure of how good the taxi-driver is at finding scarce passengers. It is the substrate concentration at which we can expect the enzyme to work at 50 per cent of its maximum rate.

Box 3 Analysing reaction rates

For the early enzymologists there was a practical problem of estimating just how high the reaction velocity would go. First of all, real-life rate measurements are inevitably subject to some degree of experimental error, but, above all, at a time when graphs were drawn by hand it was not easy to extrapolate a curve reliably. However, in the decades that followed, several ways were found to rearrange the Michaelis–Menten equation to a form predicting a straight-line graph. For example, turning both sides upside down, we get:

$$\frac{1}{v} = \frac{K_m + [S]}{V_{max}[S]} = \frac{1}{V_{max}} + \frac{K_m}{V_{max}[S]}$$

This restated form of the equation tells us that if we plot a graph of $1/v$ against $1/[S]$ we should get a straight line with a slope of K_m/V_{max} and crossing the vertical axis at $1/V_{max}$. This is the Lineweaver-Burk plot (Figure 9) and it is now very easy to extrapolate the straight line and read off the values of the two constants K_m and V_{max} (actually $1/K_m$ and $1/V_{max}$, but the conversion is trivial).

The method of analysis in Figure 9 has persisted, because it turned out to be a remarkably good description of the behaviour of one enzyme after another. Until the 1960s the analysis was done by hand, with pencil and graph paper, but computers entered biochemistry labs in the 1960s and made manual plotting of linear graphs redundant. One could now simply ask the computer to produce the best curve obeying the Michaelis–Menten equation to fit the experimental values of v and $[S]$. Various software packages will carry out the task and produce the graphs not now as an analytical tool but simply as an illustration.

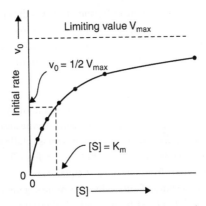

8. Dependence of reaction rate on substrate concentration with V_{max} and K_m shown. v denotes velocity (rate) of the reaction and [S] is substrate concentration.

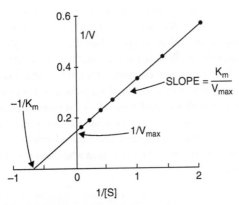

9. Lineweaver–Burk plot. With a well-chosen range of substrate concentrations the plot allows estimation of both K_m and V_{max}.

Thus, if K_m is a very low concentration, it means the enzyme is extremely good at grabbing its substrate molecules even when they are thin on the ground (or in the test tube). This analysis allows us to redraw graphs such as Figure 5(b) in the form shown in Figure 8.

It is easy to underestimate the simple concept underlying this analysis, the idea of a 1:1 complex between enzyme and substrate molecules. This *molecular recognition* model is totally taken for granted by biochemists nowadays, but it is profoundly significant because, quite apart from enzymes, it equally accounts for the way in which the molecules of antibodies, hormones, gene switches, viruses, etc., all interact with their biological targets. It accounts, in short, for the precision and selectivity of biological interactions that underpin all life processes. How this is possible should become clearer in Chapter 3.

Finally, to recapitulate, we now have a picture in which a particular enzyme will enhance the rate of a particular reaction by decreasing the activation energy. It does this by direct physical interaction between the enzyme molecule and the substrate molecule(s) in a transient complex. This interaction cannot choose or alter the direction of reaction. The direction is set by the experimental conditions and above all by the concentrations of all the reactants and products, and reaction inevitably has to proceed towards the equilibrium position. The enzyme's role is an enormous acceleration of the reaction in whichever direction it wants to go.

Chapter 3
The chemical nature of enzymes

Remarkably, the insights into enzyme catalysis outlined in Chapter 2 were achieved without any knowledge of what enzymes themselves are in chemical terms. As we have seen, that analysis suggests that enzymes act by grabbing hold of their substrate(s) in an intimate embrace. Inevitably, however, we want to know what kind of chemical structure could account for both the potency of enzyme catalysis and the very tight substrate specificity that enzymes display. In the absence of any inkling as to the nature of enzymes themselves, the tale of biocatalysis remained dangerously close to black magic, or, worse still, vitalism. Paradoxically also, the potency of enzymes made it harder to discover what they were. A tiny amount of material, contained in a complex biological soup, can be responsible for a very large effect, but a tiny amount of material is hard to find and identify, the classic needle-in-a-haystack problem. There were, nevertheless, some clues early on. Enzymes seemed, in general, to be far more fragile than typical chemical catalysts. In particular, they tended to be easily and irreversibly inactivated by heat and were also very sensitive to changes of pH (acidity or alkalinity). What kind of molecule might fit these characteristics?

Isolation of enzymes

To discover the chemical nature of these fragile agents it was essential to obtain them in a pure form, that is, to get rid of everything else. The early biochemists devised various procedures to this end. These invariably relied on the principle of *fractionation*, that is, separating a solution into two portions that behave differently. Sometimes separation was based on solubility. The solution might be heated to a chosen temperature or treated with an organic solvent or exposed to increasing concentrations of a salt. Different materials will respond to such changed conditions in different ways, but typically part of the dissolved material becomes insoluble and can then be removed by filtration or with a centrifuge, while the remainder stays in solution. In an ideal procedure, all the desired enzyme activity is in one of the two fractions while a large amount of inactive material is removed in the other fraction. In reality, separations are seldom as easy and complete as this, and fractionation procedures usually need several steps entailing substantial losses along the way. Another approach is adsorption on a solid material, usually in the form of fine particles packed to form a tall flow-through column. In the early days the solid material would be a readily available substance such as charcoal or alumina. Nowadays we have a wide range of designer materials for this purpose. Some of the dissolved material will adhere and some will remain in solution. For historic reasons, this approach is called *column chromatography* (Figure 10). This implies separation of coloured substances; enzymes are most often colourless, but we can detect them in other ways and the name persists.

By fine-tuning such fractionation steps and combining them sequentially, researchers were often able to purify their enzymes several-hundredfold or even more. In the early days, the extent of purification would be assessed by the amount of enzyme activity per unit dry weight of dissolved material, although, as we shall see, a much better basis of assessment became available in due course.

Purification and controversy

Purification of enzymes might have been expected to lead unerringly to a clear identification of their chemical nature.

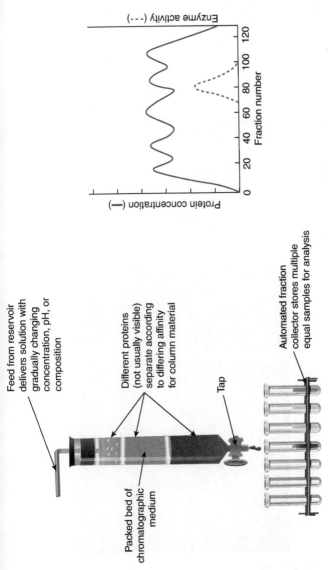

10. **Column chromatography.** Different proteins in the solution loaded onto the top of the column separate as they wash down and are collected in dozens of tubes for analysis.

Instead it led to controversy throughout the 1920s so that, even in 1930, in his milestone monograph *Enzymes*, the English scientist and prolific author J. B. S. Haldane felt obliged to sit on the fence. On one side of the argument, two American biochemists, James Sumner with jack-bean urease (Figure 11) and John H. Northrop with pancreatic trypsin, had both found that their purified enzyme material could crystallize. Solid crystals, with a precise geometric form and sharp edges, are formed when molecules line up in an orderly three-dimensional (3-D) array.

Crystallization is often used as a method of achieving and/or demonstrating purity. The fact of crystallization implies that the molecules themselves must have an orderly reproducible structure. Otherwise they could not form an orderly array. These *crystals* showed the characteristic test properties of *protein*

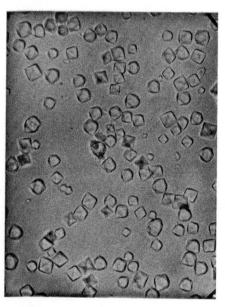

11. **Urease crystals.**

(by then established as one of the major classes of biochemical molecules and one of the critical components of food), and Sumner and Northrop therefore maintained that these protein crystals were indeed the crystals of their pure enzymes. However, other scientists were initially reluctant even to accept that a protein could crystallize, and over the years many biochemists have indeed faced the embarrassment of supposed protein crystals that turned out merely to be crystals of salt. Sumner's ideas were attacked by the eminent German Nobel Laureate Richard Willstätter, well qualified to express an opinion, having purified a number of different enzymes and made major contributions to our understanding. Specifically, Willstätter's objection rested on the fact that his own highly purified preparations of a different enzyme, peroxidase, showed no measurable protein content, and so, he argued, enzymes could not possibly be proteins.

We now know that the Americans were right; with very few exceptions (see Chapter 6), biological catalysts are in fact proteins. These are a class of chemicals that have all the fragile properties listed above, properties we see in our kitchens when we boil an egg or watch milk curdling in the presence of anything acidic. How, then, can we account for the disagreement? In theory, it might have meant that some enzymes (e.g. trypsin and urease) are proteins and others (e.g. peroxidase) are not, but this is not the explanation. Two reinforcing factors conspired to lead Willstätter to a false conclusion. First, the methods then available for detecting protein in a solution were not sufficiently sensitive. Second, peroxidase is an exceedingly potent enzyme, so that a solution containing only a minute concentration of peroxidase molecules can nevertheless exert a remarkable catalytic effect. Willstätter's potent peroxidase preparations were just too dilute for the protein to be detectable by the methods of the time.

At that stage of enzymology there were no clear criteria for deciding whether an enzyme needed to be purified a hundredfold, a thousandfold, or maybe even a millionfold in order to achieve purity. Indeed this can only be calculated once someone has achieved a degree of purity demonstrably close to 100 per cent. Today we not only know that it is always a protein we are looking for, but also we have excellent analytical methods for separating and displaying different proteins (Figure 12) and so we aim to obtain an enzyme preparation that contains only one detectable protein.

This also means that, in monitoring a purification nowadays, we focus above all on the units of enzyme activity *per unit weight of total protein*. Certainly we still want to get rid of other

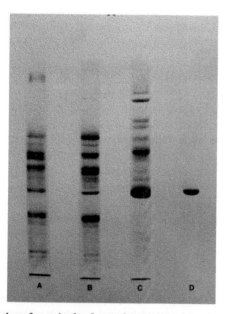

12. **Separation of proteins by electrophoresis on polyacrylamide gel in presence of detergent sodium docecyl sulfate (SDS). In this very popular method, proteins, unfolded by SDS, separate according to size, sieved by the gel. Usually invisible, they are revealed after electrophoresis by staining with dye.**

contaminants such as nucleic acids, carbohydrates, fats, etc., but we no longer suspect that any of these will turn out to be the enzyme, whereas any protein contaminant is a potential candidate. Table 1 contains figures for a typical purification procedure showing the accountancy exercise balancing purification factor against overall yield of enzyme units.

What are proteins?

If we accept that enzymes are proteins, we need to consider in more detail what proteins are chemically and whether their properties might help to explain how some proteins are able to carry out remarkable feats of catalysis. Proteins are biological *polymers*, that is, they are built of similar units joined end to end to form very large molecules. Some polymers, homopolymers, are made with identical units, but proteins are *heteropolymers*— that is, their units, though similar, are not identical. The units in this case are *alpha* - (α-) *amino acids*. Their shared, defining structure is shown below (Figure 13).

There is a central C, the α-carbon atom, and we see four 'arms' corresponding to the four valencies (Chapter 1) of carbon atoms. One arm is linked to $-NH_2$, a so-called amino group. A second arm joins the α-carbon to $-COOH$, a carboxyl group. This group confers acidic properties on a molecule. Hence the structure is that of an *amino acid*. One of the two remaining arms simply carries a hydrogen atom, H, and the last one is linked to an 'R-group'. The R is a 'could-be-anything' label: in the simplest, smallest amino acid, glycine, it is just a second H, whereas in the largest one, tryptophan, it is a bulky double ring structure (Figure 14).

In a protein molecule the amino acids are joined in a linear sequence, like railway carriages, with each amino group chemically linked to the carboxyl group of the next amino acid to

Table 1. Purification of an enzyme

Fraction	Protein (mg)	Total activity	Specific activity (total activity ÷ mg protein)	Overall purification factor	Activity yield (%)
STEP 1 Ox liver mitochondrial extract	4,165	180	0.043	1	100
STEP 2 Fractionation with ammonium sulfate	2,770	178	0.064	1.49	99
STEP 3 Active fraction from ion exchange column	296	119	0.40	9.3	66
STEP 4 Chromatofocusing	27	40	1.42	33	22

Notes: These are real data for the purification of an enzyme involved in fat oxidation. This enzyme is only present in the mitochondria (the tiny intracellular compartments responsible for trapping most of the energy from aerobic breakdown of foodstuffs). Thus at the starting point a very considerable purification has already been achieved by separating mitochondria from the rest of the cell contents. Contrast Step 2, which retains nearly all the activity but achieves only a modest purification, with Step 4, which delivers the final pure enzyme protein but at the cost of losing two-thirds of the 66 per cent activity remaining after Step 3. Chromatofocusing is an electrical method that separates proteins in a gradient of varied pH (acidity/alkalinity).

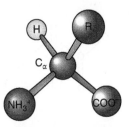

13. α-amino acid structure. The generic structure. The α-carbon carries the amino and carboxylic acid groups that give these compounds their collective name and also the sidechain 'R' group.

$$^+H_3N — C — COO^-$$

H
|
$^+H_3N — C — COO^-$
|
H
Glycine
(Gly, G)

$^+H_3N —C — COO^-$
|
H
Trytophan
(Trp, W)

14. Amino acids glycine and tryptophan. The smallest and largest of the twenty protein-forming amino acids.

form a *peptide bond* (Figure 15). In the peptide bond the carboxyl group contributes the –CO and the amino group provides the –NH. If there are, say, a hundred amino acids in a particular protein molecule, then they will be linked by ninety-nine peptide bonds. There will be just one remaining free α-amino group at one end and just one free α-carboxyl group at the other end. Chemically, therefore, a protein molecule is a linear *polypeptide*.

15. Peptide bond. The link that joins each amino acid in a protein molecule to its neighbours in the chain.

Amino acid sequence

We now know that the molecules of each individual protein have their own precise sequence of amino acid units, the *primary structure*, selected from a menu (Box 4) of twenty possible R-groups (i.e. twenty different amino acids) and dictated by the corresponding sequence of DNA bases (G, C, A, T) in the gene for that protein. Molecules of different proteins are of different length, typically 200–500 amino acid units but occasionally much longer or shorter. In 1957, Frederick Sanger (Figure 16) at Cambridge won the first of his two Nobel Prizes for devising the first working procedure to devise the exact sequence of amino acids in a protein, in his case the hormone protein *insulin* (Figure 17). This was done by systematically breaking down the protein into sets of fragments and analysing these.

Also at Cambridge, in 1953, James Watson and Francis Crick revealed the role of nucleic acids as the genetic material and we now know that the cellular machinery interprets the linear DNA coding sequence of a gene to produce a corresponding protein amino acid sequence, each amino acid being encoded by a three-letter 'codon'—for example, TTT in the DNA (meaning three *t*hymidine bases in a row) encodes the amino acid lysine. This

16. **Frederick Sanger. Sanger achieved the rare feat of winning two science Nobel Prizes, for transformative breakthrough methods for sequencing first amino acids in proteins and then nucleotides in DNA.**

coding relationship between DNA sequence and protein sequence, the *genetic code*, was worked out during the 1960s with the aid of synthetic pieces of nucleic acid with simple known sequence.

For the first twenty-five years after Sanger's pioneering work with insulin, amino acid sequences were gradually and painstakingly accumulated for individual proteins, and the corresponding DNA sequences were inferred by using the genetic code. This situation was rather rapidly reversed in the 1980s, again through the genius of Sanger, who found a quick way to solve what had seemed the impossible technical challenge of directly determining DNA sequence, earning a second Nobel Prize. The ensuing torrent of gene sequences led to amino sequences even for proteins that had not yet been found or whose true function had not yet been

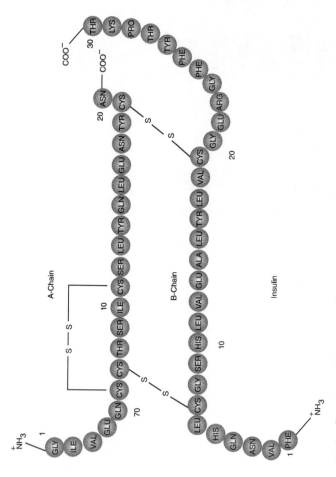

17. Insulin amino acid sequence. This hormone molecule has two polypeptide chains held together by disulfide bridges. Three-letter abbreviations as in Table 2.

Box 4 Amino acid shorthand

Protein molecules are built up from an assortment of twenty different amino acids. Some have long names, and, since we frequently want to look at long sequences of amino acids, it is convenient to have abbreviated ways to represent them. One way is a set of three-letter labels, mostly the first three letters of the full name (Ala for alanine, Leu for leucine, and so on). For very long sequences it is more compact to use a single letter for each amino acid. With twenty-six letters in the English alphabet and twenty amino acids to specify, this ought to be easy, but unfortunately those who chose the names did not foresee the problem of having several amino acid names starting with the same letter (e.g. alanine, arginine, aspartic acid, asparagine). The single-letter abbreviations are therefore not all obvious. As we shall use both types of abbreviation later in the book, they are listed in Table 2

Table 2. Standard abbreviations for amino acids

Full name	Three-letter abbreviation	Single-letter abbreviation
Alanine	Ala	A
Arginine	Arg	R
Asparagine	Asn	N
Aspartic acid	Asp	D
Cysteine	Cys	C
Glutamic acid	Glu	E
Glutamine	Gln	Q
Glycine	Gly	G
Histidine	His	H

(Continued)

Box 4 Continued

Full name	Three-letter abbreviation	Single-letter abbreviation
Isoleucine	Ile	I
Leucine	Leu	L
Lysine	Lys	K
Methionine	Met	M
Phenylalanine	Phe	F
Proline	Pro	P
Serine	Ser	S
Threonine	Thr	T
Tyrosine	Tyr	Y
Tryptophan	Trp	W
Valine	Val	V

determined. As time has gone on, DNA sequencing has been automated, miniaturized and has become very much cheaper than at the outset, and, even though methods for direct amino acid sequencing have also improved enormously, today the sequence of amino acids in a protein of interest is most often read out from an already-sequenced gene.

The shape of protein molecules

Linear sequences are fascinating in their own right and have provided ever more information about evolutionary relationships (see Chapter 6), between different proteins, between populations, between species, but this alone gives no insight into how proteins work. Protein molecules are not usually like stiff, straight sticks of dry spaghetti; they are flexible. However, neither are they like random tangles of cooked spaghetti. If they were, they would not be able to crystallize. They can crystallize because, like soldiers on a parade ground, the individual molecules of a particular protein look identical and line up in a regular array. As it turned out, the fact that proteins can crystallize was the key to solving their structure. Early in the 20th century Lawrence Bragg, at Leeds University, had shown that crystals of simple molecules like table salt, sodium chloride, scatter X-ray beams to give a 3-D pattern of 'spots' of varying intensity that could be captured on photographic plates and read backwards to deduce what must have been the structure to produce the pattern. In the 1930s yet another Cambridge scientist, the refugee Max Perutz, set out on what most scientific colleagues took to be the mad endeavour of using *X-ray crystallography* to solve the structure of the very much larger and more complex molecules of a protein. It took twenty years to crack the problem, but in the 1950s the Cambridge crystallographers triumphantly presented the solved structure of *myoglobin*, a protein that stores oxygen in the muscles of mammals. (The choice of this particular protein for study by a team led by John Kendrew relates to its abundance and also to the fact that, as proteins go, this one is small, with only 153 amino acids.) Perutz's own project was to solve the structure of a similar, related but larger protein, *haemoglobin*, which enables our red blood cells to pick up oxygen in the lungs and carry it round the body to be released where it is needed, and this also was solved in 1958.

Like insulin, neither of these proteins is an enzyme. Nevertheless they revealed two features that turn out to be true for all proteins. These are both a degree of regularity and a degree of irregularity leading to a unique shape. This property of forming a precise complex shape is the key to understanding how enzymes work. The myoglobin structure in Figure 18 is made up of a series of rod-like sections folded into a seemingly random 3-D shape. Looking closely at the rod-like sections, we see that each one is made up of a regular coil. This structure, the *α-helix*, is one of a small number of regular patterns that are readily adopted by peptide chains and are categorized as *secondary structure*. However, the molecule of myoglobin is not made up of a single

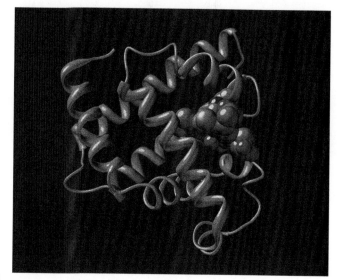

18. **Myoglobin 3-D structure. The first protein 3-D structure solved by X-ray crystallography. The model at this resolution shows the overall fold of the polypeptide chain without the detail of the individual amino acid sidechains.**

α-helix but rather of several helices of varying length joined by loops and pointing in various directions. This 3-D arrangement is the protein's *tertiary structure*. In the case of myoglobin, this third level of structure is as far as it goes, but in haemoglobin there is a fourth layer (*quaternary structure*), because four myoglobin-like subunits cluster together, like the pieces of a blackberry, to form a stable larger structure. Table 3 summarizes the hierarchy of levels of protein structure and what each level entails.

Protein folding

The solution of the first protein structures was such a mountain to climb that it is amazing, sixty years on, to note that we now have over 100,000 such protein 3-D structures in an international

Table 3. Hierarchy of levels of protein structure

Level of structure	Definition and features
Primary	Linear sequence of amino acids in the polypeptide chain
Secondary	Regular, repetitive folded structures, e.g. α-helices, β-sheets, etc.
Tertiary	Folding of secondary structure elements to form an irregular but precise 3-D object
Quaternary	Assembly of a cluster of folded polypeptides, sometimes identical, sometimes not, to form a larger functional molecule

database. They show an immense range of sizes and shapes and we know that the shape is essentially built into the linear sequence of amino acids. That is to say the newly made polypeptide chain, spooling out of the protein synthetic machinery of the cell, somehow seems to 'know', out of the vast number of possible folding arrangements, the correct one to head for and finds its way to the destination. This remarkable fact can be put to experimental test for many proteins by deliberately unfolding them to 'random coils' and then changing the conditions to see if they can find their way home. Often some molecules get tangled and lost on the way, but an impressively high percentage do find their way home, refolding to restore the biologically active 3-D structure.

Forces in protein molecules

Finally, before moving on to look more closely at enzyme catalysis, we should briefly consider the kinds of forces and interactions that occur in a protein molecule. These help us to understand both what stabilizes the constant structure of the biologically active form of a protein molecule and what kinds of interaction could steer a newly made protein molecule down the correct paths to that final folded state. Even more important for our purpose, these same interactions will help us also to understand how an enzyme molecule interacts with its substrates. These forces are not restricted to protein molecules, nor indeed more widely to biological molecules. They are forces that apply in conventional chemistry, operating also between much smaller chemical substances. The only thing that is uniquely biological is the way in which Nature, through the process of natural selection, has gradually assembled and orchestrated multiple individual interactions in 3-D space to create powerful, seemingly tailor-made solutions to the catalytic challenges of each life process.

Probably the easiest of these forces to understand is the *electrostatic effect*. This refers to the strong attraction between positively and negatively charged objects and the corresponding repulsion between things carrying the same charge, whether positive or negative. If we return to the building blocks of protein structure, the individual amino acids each have an amino group which under normal biological conditions will carry a positive charge and a carboxyl group which will carry a negative charge. However, as we saw, except for the two extreme ends of the polypeptide chain, these charges disappear in the formation of a protein. The main source of electrostatic charge in proteins, therefore, comes, not from the α-amino and -carboxyl groups, but from the sidechain R-groups that distinguish the twenty protein amino acids. Two of the twenty carry a negative charge and another three carry (or can carry) a positive charge under normal conditions in a living cell (Figure 19). A typical protein will have many of each of these amino acids in its primary sequence and therefore carries a large number of individual charges, both positive and negative, all potentially available to form favourable interactions with amino acid sidechains, or other molecules, carrying an opposite charge.

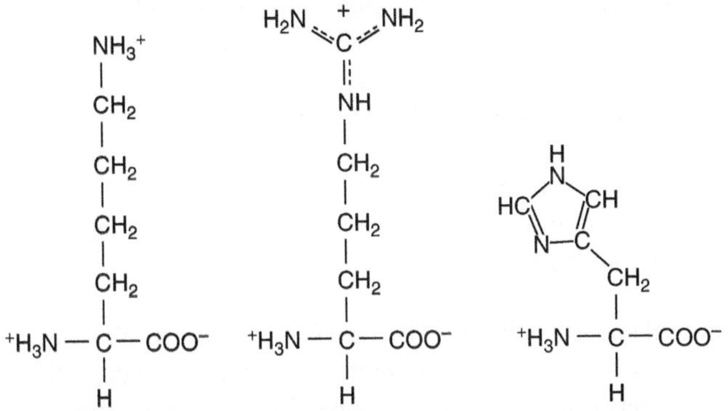

NH₃⁺
|
CH₂
|
CH₂
|
CH₂
|
CH₂
|
⁺H₃N — C — COO⁻
|
H

**Lysine
(Lys, K)**

H₂N ⸺ C⁺ ⸺ NH₂
‖
NH
|
CH₂
|
CH₂
|
CH₂
|
⁺H₃N — C — COO⁻
|
H

**Arginine
(Arg, R)**

HC ⸺ N — CH
      |
HC ⸺⸺ C
            |
            CH₂
            |
⁺H₃N — C — COO⁻
            |
            H

**Histidine
(His, H)**

The basic amino acids lysine, arginine, and histidine.

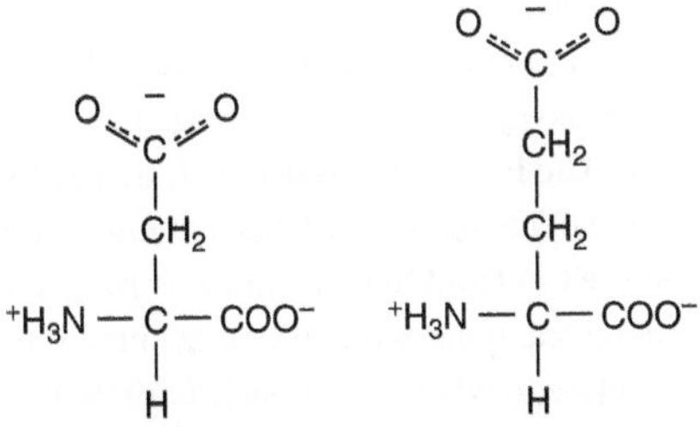

O ⸺ C⁻ ⸺ O
|
CH₂
|
⁺H₃N — C — COO⁻
|
H

**Aspartate
(Asp, D)**

O ⸺ C⁻ ⸺ O
|
CH₂
|
CH₂
|
⁺H₃N — C — COO⁻
|
H

**Glutamate
(Glu, E)**

Amino acids with side-chain
carboxylates.

19. Charged amino acids. The figure shows the five protein amino acids with sidechains that are usually/often charged.

The second major influence is that of *hydrophobic interaction*. Everyone, without knowing it, is familiar with this phenomenon, because everyone has seen how droplets of oil, suspended in water, will immediately coalesce if they meet. This is because, by doing so, they decrease the combined surface area and so minimize the energetically unfavourable interaction at the oil–water interface. Nearly half of the twenty protein amino acids have sidechains (R groups) that are predominantly hydrocarbon in nature and are correspondingly hydrophobic. When a protein is folding there is a driving force for these sidechains to huddle together, ending up in the interior of the molecule, forming the so-called *hydrophobic core* .

A third important driver is *hydrogen bond formation*. We have seen how the amino acids in a protein are joined through peptide bonds between a carbon double-bonded to oxygen and a nitrogen atom with a hydrogen attached. The C=O in a peptide bond is a potential *hydrogen bond acceptor* and the NH is a potential *hydrogen bond donor*. Their spatial relationship does not allow the C=O and the NH in a single peptide bond to interact in this way, but it is perfectly possible for the C=O of one peptide bond to approach the NH of another. If this happens, there is a tendency for the acceptor oxygen to lure the hydrogen away from the nitrogen atom. The H ends up undecided, shared between the two atoms, a chemical *ménage à trois*! The net effect is to form a weak link between the C=O and the NH, not as strong and permanent as a proper chemical bond but still strong enough to influence what ends up close to what. As far back as 1948, Linus Pauling, the visionary American chemist who won the Nobel Prize not only for Chemistry but also for Peace, spotted the fact that the regular recurrence of peptide bonds all along the polypeptide chain meant that the right way of folding would allow the formation of multiple hydrogen bonds.

If, in fact, the chain was folded into a spiral staircase, a helix, with 3.6 amino acids per turn, this allowed all the C=O and NH groups to form almost parallel hydrogen bonds, making the spindles of the staircase (Figure 20). Pauling *predicted* that such an obviously strong rod-like structure must be an important element of protein structure. This was the α-helix, and as we have already seen,

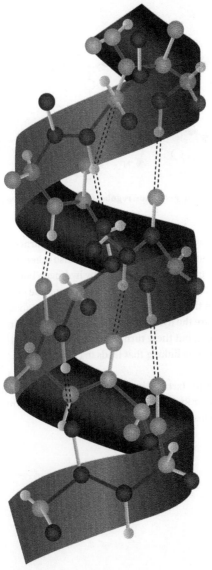

20. α-helix. This very common feature of protein structures gains its strength from multiple, approximately parallel hydrogen bonds.

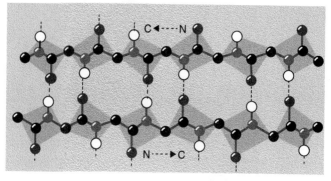

21. β-sheet. Another common regular secondary structure pattern in proteins.

almost ten years later, the myoglobin molecule turned out to be made up of multiple α-helices. There are also other stable secondary structures made possible by hydrogen bonding (e.g. Figure 21). Furthermore, not only the peptide bonds can participate in this type of interaction. Less predictable, because less regular, are the hydrogen bond partnerships of various amino acid sidechains, but these turn out to be important in the specific enzyme–substrate linking that leads to catalysis.

We need now to start putting the pieces of the jigsaw together to find out how far knowing the chemical nature of enzyme molecules helps us to understand how they act.

Chapter 4
Structure for catalysis

Structural complementarity—the lock-and-key model

One of the striking features of enzyme catalysis that we have already commented on is *substrate specificity*, that is, the fact that enzymes are exceedingly choosy about which molecules they will accept as substrates. To account for this, already in 1890, ahead of most of the mathematical analysis of reaction rates described in Chapter 2 and long before the protein nature of enzymes was revealed, the German chemist Emil Fischer had put forward the *lock-and-key hypothesis*. The enzyme is viewed as the precisely shaped lock, and only the right key, the substrate, can fit and turn it. Now that we know that enzymes are proteins, with an almost infinite range of available sequences, and hence also an endless choice of folded shapes, we can see the potential over time for Nature to evolve protein molecules that provide the ideal crevice to fit perfectly around each new substrate 'key'.

Enzyme-substrate complexes—fact or fancy?

The lock-and-key combination, of course, is the E-S complex we met in Chapter 2 in the Michaelis–Menten theory, and today this is no longer purely hypothetical. The first solved protein 3-D structures (Chapter 3) were not for enzymes, but it was not long before, in 1964, David Phillips' crystallography group in London solved the first enzyme structure. As with myoglobin, to make the task manageable, the enzyme was chosen for its unusually small molecular size. The enzyme *lysozyme* is a defence against bacterial infection, able to cut holes in the complex carbohydrate chains of the tough bacterial cell wall. It is found, amongst other places, in tears, protecting the eye, and in birds' eggs, protecting the embryo. Phillips' group solved the structure of hen egg-white lysozyme. Although this was not yet the structure of an actual E-S complex, it nevertheless displayed an obvious groove across the enzyme surface (Figure 22) long enough to accommodate six sugar units of a carbohydrate chain. This immediately led on to molecular modelling exercises to discover the optimal snug fit, which, in turn, made it possible to formulate a plausible chemical mechanism of catalysis involving the side chains of Glu 35 and Asp 52.

Ideally, for any enzyme, solving the structure of the E-S complex itself would eliminate the guesswork and doubt associated with the modelling approach, but, for an enzyme that works with a single substrate, there is the obvious problem that the target will not stand still! As soon as enzyme and substrate meet, the catalysed reaction will take place. There are two possible solutions to this problem. The first is to use additives, allowing the solution to be cooled well below the normal freezing point. This may slow down even a catalytic reaction sufficiently to allow crystallographers to capture the necessary data before the complex reacts and breaks down. This has been made possible by the use of very intense X-ray sources and very rapid data capture devices. The second possible trick is to use the equivalent of a blank cartridge in a gun—that is, a molecule that is so similar to the true substrate that it fits the same attachment site on the enzyme surface but nevertheless lacks the key feature to undergo reaction. Such an impostor compound is referred to as a *competitive inhibitor* because under normal reaction conditions it will indeed compete with the true substrate for the available slots on the

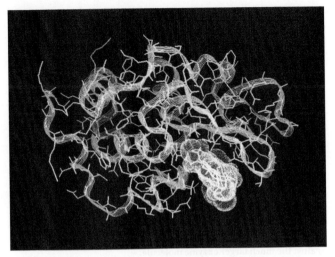

22. Lysozyme structure with modelled substrate. The first solved enzyme structure.

enzyme surface and will therefore decrease the rate of the catalytic reaction. In the context of crystallography, using a competitive inhibitor instead of the substrate allows formation of an enzyme-inhibitor complex that resembles the true E-S complex but is stable. This approach is open to the objection that it might only deliver partial truth, because there could be subtle differences induced in the enzyme protein when the authentic substrate molecule is present, but it at least shows unambiguously where a substrate-like molecule sits on the enzyme surface.

Very many enzymes, however, catalyse reactions involving more than one substrate, and in these cases the full catalytic drama cannot occur without the full cast of actors. For such enzymes, therefore, if a complex can be formed with a single substrate in the absence of the other reaction partner(s), it should be stable and thus allow crystallography of an authentic E-S complex without any elaborate tricks. Many such complexes have been studied, offering insight into the details of molecular recognition (Figure 23).

23. Solved structure of an E-S complex. The international protein database now contains many structures for E-S complexes. This one, for human haem oxygenase, is suitable for illustration because the size and shape of haem, with four rings joined in a square, makes it visible against the much larger enzyme molecule.

Crystallographers are obliged to deposit the 3-D coordinates for their solved structures in an international protein database (https://www.wwpdb.org/). This information is freely available to anyone interested and is usually at *atomic resolution*—that is, it specifies the position of all, or at least the majority, of the atoms in the structure so that one can clearly see the shapes of individual amino acid sidechains and see in an E-S complex exactly how the substrate is anchored.

Catalytic groups

We can thus now not only infer, like the early biochemists, that enzymes must be able to fit their substrates like a glove fits a hand, but also in many cases see the direct structural evidence. However, in itself, this still does not explain enzyme action. After all, a hand goes in and out of a glove many times and still remains a hand complete with all its fingers. A crucial ingredient of the enzyme's equipment for achieving outstanding catalysis is that, as well as providing ideal anchorage points for its substrate(s), the so-called *active site* on the enzyme surface will also have a few 'catalytic groups' ideally positioned to interact with the portion of the substrate molecule that is to react. These catalytic groups do the chemistry and are selected from the repertoire of twenty different kinds of amino acid sidechain. They include, for example, the carboxylate (-COO⁻) groups of aspartic and glutamic acid, the imidazole ring of histidine, the amino group (-NH$_3^+$) of lysine, the thiol (-SH) group of cysteine. These are, in principle, no different from the kinds of catalyst a chemist might use in the lab, but the critical difference is that, in place of randomly orientated chance collisions in solution, the enzyme provides perfect spacing and orientation. It not only ensures that the necessary encounter(s) take(s) place but also locks in the ideal geometry, so that the encounter has the optimal chance of success—that is, of bringing about the chemical reaction. In recent years chemists have attempted to learn from Nature, synthesizing 'smart' catalysts that utilize the same principles of precise anchorage and orientation. This is known as *biomimetic catalysis* and, not surprisingly, has proved to be very successful.

The importance of flexibility

An important further dimension to our understanding came with the realization that enzyme molecules are not necessarily rigid. In the 1950s, examples started to emerge of molecules that seemed to fit the lock in the old Emil Fischer model and nevertheless failed to turn it. It is easy to see why a molecule that is too big cannot fit the lock, but if a molecule is smaller than the 'official' substrate and nevertheless still has all the atoms involved in the chemical reaction, why should it not work as a substrate for the enzyme? This led to the concept, formalized by Daniel Koshland at Berkeley, of 'induced fit', according to which an extra bit on the substrate molecule provokes a *flexible* enzyme molecule to undergo a subtle movement, perhaps enfolding the substrate more closely or bringing a key catalytic group into proximity and alignment. Further insight into flexibility came from various enzymes that catalyse two-substrate reactions, for example the reaction mentioned in Chapter 2 that produces lactic acid in our muscles. *Lactate dehydrogenase* catalyses a reversible reaction that either oxidizes lactic acid to produce pyruvic acid or reduces pyruvic acid to produce lactic acid. This reaction requires the *coenzyme* NAD$^+$ (see later in this chapter for a fuller explanation):

$$\text{Lactic acid} + \text{NAD}^+ \leftrightarrow \text{Pyruvic acid} + \text{NADH} + \text{H}^+$$

Analysis of this reaction proves very convincingly that the enzyme follows a *compulsory-order* mechanism in which the coenzyme NAD$^+$ has to come onto the enzyme before lactic acid (and similarly for NADH in the opposite direction of reaction). One might have assumed that an enzyme highly specific for lactic acid would have a perfectly fitting site waiting to accommodate that substrate, but in effect there is no site for lactic acid until after the NAD$^+$ has arrived. Crystallographic studies in the laboratory of Michael Rossmann in Indiana showed beautifully that, upon arrival of the coenzyme, a large fourteen-amino-acid loop folds in to enclose the active site, assembling the attachment points for lactic acid or pyruvic acid.

With induced fit, we are now invoking action of the substrate on the enzyme, but already in the 1930s and 1940s people like J. B. S. Haldane and Linus Pauling had talked about the enzyme imposing *strain* on the substrate molecule. They saw this as an important ingredient of the enzyme's catalytic action. Superficially these may appear to be two opposite ideas, but, in truth, interaction between any two bodies, big or small, is a two-way process. Neither is totally rigid and, if the enzyme is stretching the substrate, so also the substrate must be stretching the enzyme. Back in Chapter 2 we met the idea of a molecule needing to be given activation energy, the necessary nudge to shift it over the lip of the hollow. Now we can see that in an E-S complex, the interaction with the enzyme protein might twist or stretch or tickle our substrate molecule just enough to destabilize it and 'tip it over the edge'.

Transition state analogues

Following on the heels of the induced-fit concept, this kind of thinking led to an important new insight. We have already introduced the idea of competitive inhibitors which strongly resemble the substrate or product of an enzyme reaction. However, in the course of a chemical reaction, there has to be a halfway house, a state in which the chemical bonds to be broken are not quite gone and the new bonds to be made are not fully formed. In this *transition state*, the molecule is neither substrate nor product but instead is something in-between with distinctive geometry. A number of scientists therefore hypothesized that an ideal enzyme structure would provide perfect tight binding, not for the substrate, nor for the product, but rather for the transition state. Such an enzyme molecule would tend to coax the substrate to the edge of the cliff, precipitating rapid reaction. With this in mind, chemically adept enzymologists set about designing *transition-state analogues* for various enzymes—that is, small molecules that faithfully mimicked the shape of the true transition state. These, rather than direct analogues of the substrate or product, they postulated, should be the tightest and most potent inhibitors. A number of these designer compounds turned out indeed to be very potent inhibitors, attaching to their enzyme perhaps a hundred or a thousand times more tightly than the natural substrate/product (Figure 24). As we shall see in Chapter 7, inhibitor design is a fertile area for pharmaceutical chemists, but in our immediate context, the success of this concept provides powerful evidence for the underlying view of how an enzyme works.

Assembling the cast

One further aspect of enzyme action to be considered is the way in which they handle reactions involving more than one substrate. Single-substrate enzymes convert A to B or maybe C to D + E, but

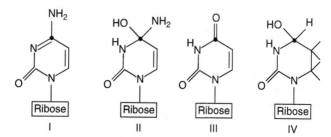

24. Transition-state substrate analogue. The enzyme cytidine deaminase removes the –NH₂ group from I (cytidine) as ammonia and forming III (uridine). This is assumed to go via II. In this transition state we have a tetrahedral C at the position marked with a black dot, unlike I and III where the C is planar. Accordingly the analogue IV attaches to the enzyme 10,000 times more tightly than the substrate (I) or product (III).

the majority of enzyme reactions involve two or three substrates and occasionally even four. These multisubstrate enzymes handle their task in one or other of two quite distinct ways which we may view as the 'marriage broker' method or the 'pigeonhole' method. The first approach involves bringing together all the participants in the same place, the enzyme's active site, at the same time (as happens with lactate dehydrogenase; Figure 25). Especially for reactions with three or more substrates, this must enormously increase the statistical chances of reaction as compared with the random chance of all the necessary molecules colliding in solution. As usual, the enzyme also imposes the correct geometry.

In the alternative pigeonhole approach, the substrates do not need to meet at all. The first substrate leaves behind a chemical 'package' on the enzyme for the second substrate to pick up in its own time. This involves a chemical reaction between the first substrate and the enzyme, producing the first product and a chemically altered protein which can in turn react with the second substrate to form the second product (Figure 26). At first sight this seems to violate our early definition of a catalyst as an agent that speeds up a reaction without itself

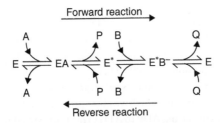

25. **Ternary complex mechanism.** In this mechanism the reactants have to meet at the active site. In the version illustrated they have to arrive in a compulsory sequence.

26. **Enzyme substitution (or ping-pong) mechanism.** In this mechanism reactants A and B never meet on the surface of the enzyme.

being altered. However, the reaction is now made up of two 'half-reactions'. Actual catalysis only occurs when you put the two half-reactions together and the enzyme returns to its original unaltered state and accordingly is able to continue with further multiple cycles of reaction. If you only add the first substrate, you would indeed have a chemically altered enzyme, but you would have no catalysis!

Enzymes' little helpers

So far we have implied that the mighty enzyme protein is self-sufficient, providing all the necessary chemistry via its amino acid sidechains and the environment offered by the folded structure. This is true for many enzymes, but quite early in the development of biochemistry it was noted that certain reactions seemed to need two components for their catalysis. These could be distinguished by their ability to cross a semi-permeable membrane and by their resistance to heating. Enzyme proteins have very large molecules that do not cross semi-permeable membranes but some of them also require small molecules that easily pass through a membrane. Enzyme proteins typically are inactivated and coagulate upon boiling, leaving behind more stable, small molecules which retain their potency. These small molecules are referred to as *coenzymes* or *cofactors*.

When these reactions were analysed more thoroughly, it turned out that some of the cofactors are actually enzyme substrates themselves, for example adenosine diphosphate (ADP), and adenosine triphosphate (ATP). The reactions that break down glucose in our cells in a process called *glycolysis* trap chemical energy in a useable form to drive other processes, and this happens by converting ADP to ATP. (This is discussed in greater detail in Chapter 5.) These reactions can be carried out in the test-tube by extracts of broken yeast cells (see Chapter 1), muscle cells, etc. Without ADP and ATP, however, several of the reactions in the linear reaction sequence of glycolysis cannot take place and the whole process grinds to a halt. The requirement for these and other cofactors was originally demonstrated by putting cell extracts capable of glycolysis (e.g. yeast extract) into a membrane sac which was tied off at both ends and washed on the outside with large volumes of liquid (Figure 27). In this process of *dialysis* (probably more familiar in the context of its medical application to treatment of patients with kidney failure) small-molecule cofactors like ADP and ATP pass through the membrane and are washed away, while the large enzyme proteins are left behind in the sac. The enzymes in the sac are unaltered and should still be able to perform. In practice, though, they cannot work without the cofactors (demonstrated simply by testing a sample). However,

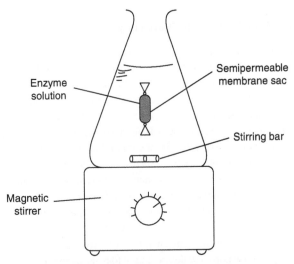

27. **Dialysis.** Typically the solution in the flask would be changed repeatedly. If the volume in the flask is 1,000 times the volume in the sac, three cycles of dialysis should decrease the concentration of low-molecular weight solutes in the sac by a factor of 10^9.

once the cofactors are replaced, the metabolic pathway immediately springs back to life.

In other cases the cofactor is an integral part of the enzyme. It may or may not be chemically attached to the protein but is in any case rather tightly bound (Box 5).

One further kind of cofactor also explains some of our trace dietary requirements—metal ions. Different enzymes require a variety of metal ions, sometimes merely to stabilize their folded structure, but often also as key components in the catalytic machinery at the active site, where they can be clearly seen in the crystallographic structures. These essential metals include iron, copper, zinc, magnesium, manganese, molybdenum, calcium, and in each case the enzymes exploit particular properties of these metal ions in order to carry out biological catalysis.

Box 5 Cofactors and vitamin requirements

Our need for enzyme cofactors accounts for many of our vitamin requirements. Thus, on the side of cereal packets, you will see listed the content of thiamine (Vitamin B1), riboflavin (Vitamin B2), pyridoxine (Vitamin B6), nicotinamide (Vitamin B3), pantothenic acid (Vitamin B5), etc. These are all organic compounds that we cannot make for ourselves but are absolutely essential for life as precursors for enzyme cofactors. We rely on more self-sufficient organisms, such as fungi and bacteria, to make them for us. Thiamine pyrophosphate is the tightly bound cofactor in enzymes that catalyse oxidative decarboxylation, crucial in our energy metabolism. Riboflavin is the precursor for FAD and FMN, and nicotinamide similarly for NAD^+ and $NADP^+$; these four are cofactors for a number of enzymes catalysing oxidation reactions. Pyridoxine is used to make pyridoxal phosphate which is linked to a number of enzymes that handle incorporation or removal of amino or carboxyl groups. Pantothenic acid leads to Coenzyme A, an essential cofactor in fat metabolism, glucose metabolism, and various other processes. These and a number of other such accessory molecules extend the chemical repertoire available to our enzymes. Depriving human beings of one or more of the vitamin precursors leads progressively to diseases such as beri-beri (B1 deficiency) and pellagra (B3 deficiency). The symptoms of these diseases are the result of progressive failure of the enzymes that absolutely rely on their cofactors. Although many of these diseases had been known for hundreds of years and were widespread well into the 20th century it was only in the 1930s that the various vitamins were discovered and their relation to enzyme cofactors revealed. Sir Frederick Gowland Hopkins, mentioned in Chapter 1, was one of the leading pioneers in this work. Even now, however, we know remarkably little about the detailed basis of deficiency symptoms. Partly this is because poverty and famine or other causes of malnutrition tend to result in multiple deficiencies, but also it is because we have paid insufficient attention to the way in which different organs and different enzymes divide up scarce cofactors when there is not enough to go round.

Catalytic power

As we have seen in this chapter, there are various patterns of enzyme mechanism and also a variety of structural features helping to achieve the objective. It is, however, worth emphasizing the remarkable outcome. Returning to our earlier example of rate enhancement by carbonic anhydrase (Chapter 2) we can reconsider the ten-millionfold acceleration now in terms of the individual enzyme molecule. If we think of each carbonic anhydrase molecule as a tiny molecular machine, it turns out its product approximately one million times per second! This is a degree of diligent efficiency that is quite difficult to comprehend when compared with typical speeds of human action or even of our own constructed machines.

Chapter 5
Enzymes in action

Literally every life process depends on the action of enzymes. Clearly, we can only examine a tiny selection of the vast range. This chapter explores a few interesting examples to illustrate the sophistication and subtlety of biology's catalysts.

Proteinases in digestion

Proteins, as we have seen, are large, varied, complex functional molecules. Living organisms constantly make them afresh from amino acid building blocks, for growth, for repair, for adaptive responses, for reproduction. The building blocks are a valuable resource, especially for creatures like ourselves, able to make only about half of the necessary twenty amino acids for ourselves and therefore reliant on food for the rest. Our food contains proteins made by other living things, and one function of the digestive system is to dismantle these, releasing the amino acids for our own use. For this we have a set of enzymes called *proteinases* which work by splitting the peptide bonds (Figure 28). Why a set? Could not a single enzyme recognize the peptide bonds and get to work? We need to imagine the reality of the task confronting a proteinase molecule. It is itself a large protein molecule attempting to get its 'teeth' into another large protein molecule. But where are the peptide bonds to be cut? They are everywhere but nowhere to be seen. Most are bundled away out of reach in the middle of the

28. **Cleavage of a peptide bond.**

folded protein, and the rest, on the surface of the substrate protein, are hiding behind the bristling and varied amino acid sidechains.

One thing that might help to make the peptide bonds a bit more accessible would be to disrupt the folded structure of the food proteins. Modern humans help this process by cooking their food. Anyone who has boiled or scrambled an egg or fried a steak will be aware that something fairly dramatic happens. However, other animals do not know how to cook, and we ourselves presumably survived for thousands of years without doing so. Within our own bodies we unfold proteins in two ways. First of all, the stomach maintains remarkably acidic conditions, down at pH 2, and most protein structures cannot tolerate this. Second, further down the digestive tract, bile pours into the mixture, adding natural detergents which also tend to unfold proteins.

An immediate problem however: since proteinases are proteins themselves, how do they remain folded? Structural studies reveal that they are unusually well supplied with internal chemical ties (disulfide bridges between cysteine sidechains) linking neighbouring lengths of the chain and thus holding the whole folded structure together (Figure 29). We can show in test-tube experiments that they are indeed remarkably resistant to unfolding.

Even with rough treatment in the gut, however, the food proteins will only partially unfold. The degrading enzymes need to find something to recognize and hold onto, and Nature has come up with a team of proteinases with different recognition patterns. The first attack, from *pepsin*, working in the acid stomach, preferentially cuts peptide bonds between amino acids with large, bulky sidechains (e.g. Figure 14). There are enough such pairs for most proteins to be broken down to a number of fairly large peptide fragments. As the mixture passes on down into the small

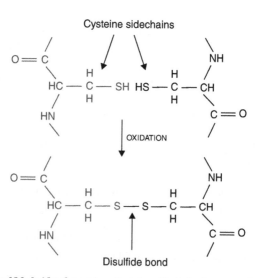

29. **Disulfide bridge formation. Cysteine sidechains have reactive –SH groups. If two of these come together they are susceptible to oxidation, forming a chemical link.**

intestine it meets the bile and the output of the pancreas, and in the mixture the acid is neutralized, resulting in mildly alkaline conditions. The pancreas contributes a set of different proteinases. Some (*exopeptidases*) nibble their way in from either end of a peptide chain, clipping off one amino acid at a time and shortening the substrate peptide. Others (*endopeptidases*), similarly to pepsin, cut peptides internally into smaller peptides. Among these are *trypsin*, *chymotrypsin*, and *elastase*. Trypsin will only cut at peptide bonds where the –CO of the peptide bond is contributed by a positively charged amino acid (lysine or arginine). Chymotrypsin requires the amino acid in that position to be large and hydrophobic. Elastase needs both partners in the peptide bond to be very small amino acids (alanine or glycine). In addition there are peptidases that specialize in splitting very short peptides. Although each of these enzymes is fussy in its requirements, acting in concert, this cocktail of enzymes is able to break down proteins entirely to their constituent amino acids to be absorbed and used.

Switching on: zymogens

This set-up sounds efficient and satisfactory, but we have overlooked a potential problem. Proteinases are dangerous. Without suitable precautions we could digest ourselves from within. Even if we avoid this fate, might not the highly efficient proteinases digest one another or even themselves in a type of enzyme cannibalism? Trypsin, for example, is indeed capable of *autodigestion*, if it has nothing better to do, a property that incidentally complicates the task of studying it on its own.

Nature's solution first of all uses the principle of 'just-in-time delivery'. At mealtime the nervous system sends a signal to the stomach to release a hormone, *gastrin*. This in turn instructs the stomach's lining cells to release hydrochloric acid and a protein, *pepsinogen*. This protein is an example of a *zymogen*, an *inactive enzyme precursor* that is only 'switched on' at the appropriate time. In this case the inactive pepsinogen molecule has 371 amino acids, but, in the very acidic conditions of the active stomach, some of it begins to activate by spontaneously shedding a 44-amino-acid piece from the N-terminal end. This releases the fully active pepsin molecule. As soon as some active pepsin appears it rapidly activates all the remaining pepsinogen. Thus the potent protein degrading activity appears when, and only when, the food is there to keep it fully occupied in gainful activity.

In the small intestine, similar mechanisms operate, but on a more ambitious scale as more proteinases are involved. The zymogens are stored away safely in granules within the cells of the pancreas. Nerve stimulation leads to coordinate release (a) of the contents of these granules into the pancreatic secretion; (b) of a special proteinase, *enteropeptidase*, produced by the lining cells of the small intestine. This enzyme controls the master switch. It has the very specific task of cutting just one peptide bond in the newly arrived trypsin precursor, *trypsinogen*, removing a 6-amino-acid peptide from the N-terminal end. The activated trypsin can now not only complete the task of processing all the remaining trypsinogen molecules, but also similarly activates several other proteinases from their precursors, including chymotrypsin, elastase, the exopeptidase carboxypeptidase, lipase (fat-digesting enzyme), etc. This is only one of a number of biological examples of *amplification*, where a small initial proteinase activity (in this case enteropeptidase) lets loose a cascade of much greater activity.

Stomachs of the cell

Breakdown of proteins into their constituent amino acids has a role not only at the level of food input for the entire organism but also, in miniature, at the level of individual cells. This is necessary for several reasons. First, enzyme molecules and other cellular structures can become damaged or wear out, and the defunct molecules need to be cleared away and recycled. Second, many tissues adapt to different challenges or physiological states and so they need both to be able to produce new enzyme teams for a new challenge and also to recycle working enzymes that are no longer needed. For example, in a pregnant female animal, the mammary gland undergoes profound metabolic changes in preparation for milk production and then again in the opposite direction at the time of weaning. Enzymes for the task need first to be produced and then eventually to be retired. Some cells also engulf external molecules or larger objects, such as bacteria, and break them down.

To deal with these matters of economy and hygiene, cells have two distinct systems, each with its own set of proteinases, but once again their destructive potential has to be controlled. In one system, the enzymes are locked away in separate structures in the cell called *lysosomes*. Like the stomach, these tiny bags of enzyme maintain a much lower pH value than the surrounding cell. Keeping the proteinases in lysosomes achieves two objectives. It means that the current working enzyme machinery of the cell is not exposed to destruction, and it also allows the cell to select and label just those molecules or other structures that need to be sent for recycling. Like the gastrointestinal digestive system, lysosomes have enzymes to break down not only proteins but also other biological structures.

The second system for degradation, a distinctive feature of animal cells, is specific for proteins and accordingly is called the *proteasome* (Figure 30). Just as in intestinal digestion, the proteasome deploys three proteinases with different specificities for peptide bonds to be cleaved. This, however, is set up like an elaborate industrial conveyor-belt tool, a single, giant, multienzyme molecule. The three proteinases face inward in a tubular structure. The proteins to be degraded are first labelled for destruction by tagging with a small protein called *ubiquitin*. One end of the proteasome tube recognizes the label, pulls the protein in, unfolding it and releasing the ubiquitin to be reused. The unfolded protein, fed down through the tube, now encounters

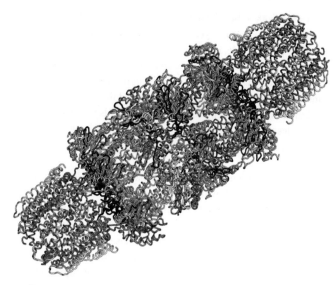

30. Proteasome.

the combined might of the three proteinases. Ultimately the small peptide fragments are released at the other end of the tube.

Cell death

Most readers will be in a fairly stable state with regard to their body size and shape and the distribution of their various organs, but this is not the case throughout our lives. As we develop in the womb and later throughout infancy and childhood, we are constantly growing and developing. To do this smoothly, we need not only to keep adding new material and structures, but also to break down and remove earlier material. Consider, for example, a leg bone maintaining its overall shape and function but increasing in size manyfold through childhood. These processes are even more dramatic in some other animals, for example a caterpillar changing into a butterfly or a tadpole into a frog. Accordingly, we need a well-regulated system not just for growth but also for remodelling. In practice this requires organized death of some of our cells and recycling of their contents. This programmed cell death is called *apoptosis*, a label that distinguishes the process clearly from *necrosis* which describes the unhealthy, rotting breakdown of cells and tissues associated with disease.

Apoptosis is a complex process, but at its heart is a cellular suicide system that is kept in check so long as the cell keeps receiving reassuring signals that it is still useful. Much of our detailed knowledge comes from studies with a 'model system', the small nematode worm, *Caenorhabditis elegans*, in which the adult has 959 cells in total, each with a tightly defined location and function (Figure 31). Like ourselves, each new worm starts as a single cell and develops through serial cell division. However, as one cell becomes two, then four, eight, sixteen, and so on, gradually the

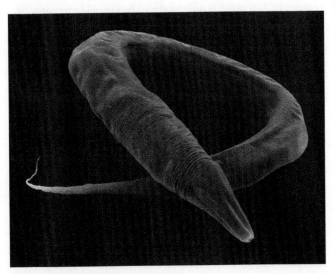

31. *Caenorhabditis elegans.*

total is whittled down by programmed cell death at various points along the way. In all, 131 cells are eliminated on the way to the final adult total.

Genetic studies of mutations affecting this process have revealed a mechanism that operates in a fundamentally similar way in all multicellular animals. Key players are a group of proteinases called *caspases* (a name coined to reflect the facts that they have a Cysteine amino acid essential to their catalysis and that they cut peptide bonds on the carboxyl side of ASPartic acid in a protein chain). In *C. elegans* the central executioner is caspase CED-3, but CED-3 remains asleep until aroused by CED-4. This is strikingly similar to the unleashing of chymotrypsin by enteropeptidase. It means, however, that CED-4 itself needs to be held in check until the time comes, and this is achieved by tying it down in a close molecular embrace with another protein, CED-9. The key action releasing CED-4 from this partnership is taken by yet another protein, EGL-1, which is only produced by the cell in response to external 'death signals'. Thus, in the development of *C. elegans*, 131 cells receive the message and switch on their own destruction at the hands of proteinase CED-3, contributing to the emergence of the final, fully sculpted adult.

Blood clotting

Our final example of the tight control of useful but potentially lethal proteinases comes from our blood circulation. As we all know, blood is vigorously pumped round our bodies non-stop, and, as a result, there is always the risk that, with a serious wound, blood will gush out and we may bleed to death. Our natural defence is blood clotting. A scratch or a puncture may bleed profusely at first, but normally, after a few minutes, a clot forms and then dries into a crust, stopping the bleeding and keeping out infection. This process is initiated and controlled by a cascade of proteinases, each one activating the next one in line. The last enzyme activation in the sequence converts inactive prothrombin to active *thrombin*. This proteinase is extremely choosy as to where it will cut, requiring a peptide bond between the amino acids arginine and glycine, and also being quite fussy about the nearby amino acids. It finds four acceptable bonds to cut in the molecule of *fibrinogen*, and cutting these four releases fibrin. Fibrin molecules then self-assemble in large numbers to form the clot.

It is critically important that clotting should be triggered very fast when there is an injury, but this very sensitive system can easily go wrong. First of all, since enzymes are encoded by genes, genetic defects in the individual proteinases lead to defective clotting and bleeding diseases. A famous example is the *haemophilia* in the descendants of Queen Victoria, including Alexis, son of the last Tsar of Russia (Figure 32). In affected individuals even a small cut, harmless in most people, can be life threatening. Nowadays one of the functions of blood banks is to purify the various proteinase clotting factors from unused blood so that they can be administered to haemophiliac patients.

In addition to the risk of clotting failing to occur promptly, there is also the opposite risk of clotting at the wrong time or in the wrong place. *Thrombosis*, that is, the formation of an inappropriate clot, can also kill us; in the coronary circulation it leads to heart attack, in the brain to a stroke. Predictably, Nature has also built in its own mechanisms to prevent the powerful clotting cascade getting out of control. There are several components to this control, but one worth highlighting is tPA, *tissue plasminogen activator*. As you may now guess, tPA is a proteinase that converts inactive plasminogen to active plasmin, itself a proteinase which attacks fibrin clots. Human tPA is now available as a drug, and its prompt administration has greatly improved the outcome for stroke victims.

Trapping useable chemical energy

Next we leave proteinases to consider how enzymes make life possible by selecting the reactions that do and do not occur and so

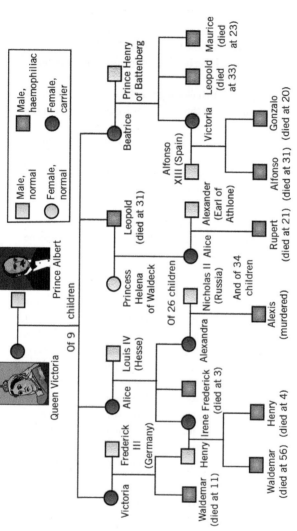

32. Royal haemophilia. The relevant gene is on the X chromosome. Queen Victoria, with one good copy and one defective copy, was a carrier, like some of her daughters, was unaffected but was a carrier, like some of her daughters. Males have only one X chromosome and Prince Leopold, inheriting the bad copy, was a haemophiliac.

directing the flow of chemicals through particular metabolic pathways. As already mentioned in Chapter 4, ATP is a central compound in bioenergetics. It is often referred to as 'the universal energy currency' of living cells. Plants and many simpler organisms exploit the energy of sunlight to drive conversion of ADP to ATP, by adding on an extra phosphate group (Figure 33). Other organisms, including animals like ourselves, drive the same conversion by breaking down foodstuffs. All organisms use the reverse conversion, of ATP to ADP and phosphate, to drive other essential life processes, both chemical ones and physical ones such as muscle contraction, nerve conduction, pumping of substances, etc.

The formation of ATP to conserve energy involves the conversion below:

$$ADP + phosphate \rightarrow ATP + water \qquad (1)$$

However, whilst this is the *net* conversion, it does not occur as a single reaction. Indeed, under any realistic set of conditions, the reaction from left to right as shown, making ATP, is impossible

In ADP, adenosine diphosphate, the circled third phosphate unit is missing.

ATP (Adenosine triphosphate)

33. Adenosine triphosphate (ATP).

because it is thermodynamically 'uphill'. As we shall see, it is precisely for this reason that ATP is useful. In order to make it in the cell, we need a phosphate donor compound that can react with ADP in an energetically feasible, 'downhill' reaction. There are various viable solutions to the problem and we shall examine just one to illustrate the principle of *energy coupling*.

In its breakdown by glycolysis (see earlier), 6-carbon glucose is converted to a sugar phosphate, with a phosphate group at either end of the molecule. The enzyme *aldolase* splits this molecule in the middle, producing two 3-carbon molecules, each with a single phosphate (Figure 34), and another enzyme, *triose phosphate isomerase*, interconverts these two so that a single downstream set of reactions can handle both.

In the next phase, the 3-carbon *glyceraldehyde 3-phosphate* is oxidized (by NAD$^+$, one of the cofactors introduced in Chapter 4). From a chemical/thermodynamic standpoint, the easy option, in theory, would be just to oxidize glyceraldehyde 3-phosphate to glyceric acid 3-phosphate, a 'favourable' reaction that could release a lot of energy. However, this is exactly what we do not want; once the energy is dissipated it is impossible to gather it up again to do useful work for us. The enzyme Nature has evolved catalyses a different reaction (2) which, as well as carrying out an oxidation, also introduces a second phosphate (from inorganic phosphate ions in solution in the cell) to give us 1,3-bisphosphoglyceric acid.

Enzymes

Glyceraldehyde 3-phosphate + NAD$^+$ + phosphate

$\rightarrow$ 1,3-bisphosphoglyceric acid + NADH   (2)

This reaction is not as steeply downhill as the theoretical one without the addition of phosphate, but is still sufficiently favourable to work. We 'waste' sufficient energy to ensure the reaction will go while keeping back some energy to store away.

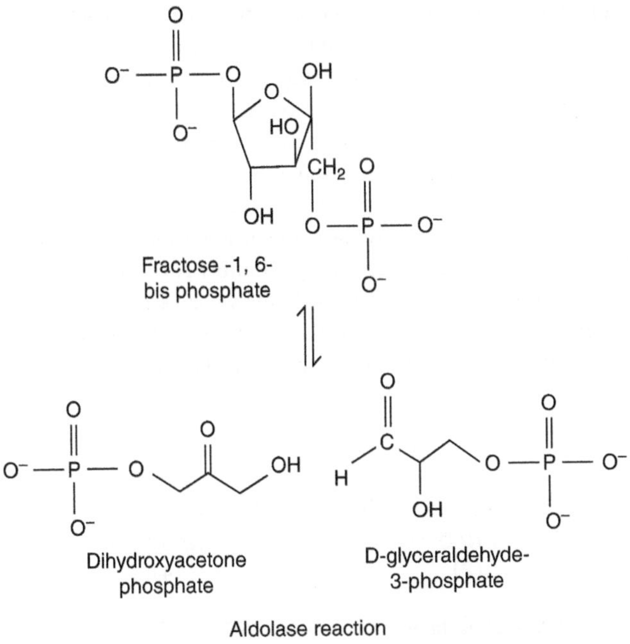

Fractose -1, 6-
bis phosphate

Dihydroxyacetone
phosphate

D-glyceraldehyde-
3-phosphate

Aldolase reaction

34. Aldolase reaction.

The brilliance of the seemingly elaborate detour, so annoying to struggling first-year biochemistry students, is that it now gives a compound that packs sufficient thermodynamic punch to be able to deliver its extra phosphate group to ADP in the very next reaction (3).

$$\text{1,3-bisphosphoglyceric acid} + \text{ADP} \rightarrow \text{3-phosphoglyceric acid} + \text{ATP} \qquad (3)$$

In Figure 35 we see two pairs of reactions. Superficially, in both cases the net outcome (the sum of both reactions) is the same and ATP is produced. However, in the first case the net outcome cannot be achieved, because one of the two reactions is not thermodynamically feasible and the other reaction squanders the energy that could have been harvested. Both reactions in the second pair are feasible, 'downhill' reactions, and one is able to drive the other, because they now have a *common intermediate* (1,3-bisphosphoglyceric acid) produced by the first reaction and used by the second. This is a little like a relay race, in which Runner No. 2 has to receive the baton in a direct handover from Runner No. 1 in order to get going. If Runner No. 1 drops the baton, Runner No. 2 has no hope. In the second scenario in Figure 35 the common intermediate 1,3-bisphosphoglyceric acid is the baton. In the first scenario there is no baton.

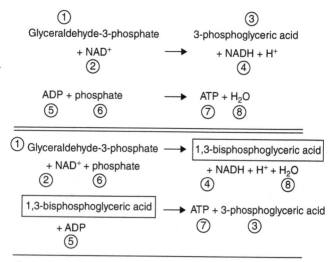

For both pairs of reactions the sum total is:

①+②+⑤+⑥ ⟶ ③+④+⑦+⑧

but only the second pair has a common intermediate so that the first can drive the second.

35. **Energy coupling via a common intermediate.**

Enzymes

This energy harvesting procedure is entirely dependent on the existence of the enzymes to catalyse the right reactions and crucially also on the absence of any catalysis of the 'favourable' reaction mentioned above which would short-circuit the whole process and ensure no ATP was made.

When it comes to using ATP to drive any one of a large number of biological processes, the same principle operates in reverse. Running Reaction (1) above from right to left should be 'downhill'. In theory it should proceed from right to left, regenerating ADP, until nearly all the ATP is split, and in the process it would release energy. Pursuing the currency analogy, this is like throwing away your money. All the ATP is gone, all the energy is released, and all the processes waiting to be driven are at a standstill. Instead of this, in every case the ATP reacts, not with water but rather with one of the components in the process to be driven so that once again there is a common intermediate and energy coupling is achieved. We shall see an example of this in the next section.

Translating the genetic code

Our next example is not one enzyme but a set of enzymes. Thanks to James Watson and Francis Crick's insight in the 1950s, we know that genetic information is encoded in DNA. DNA molecules are made up of lengthy linear sequences of *nucleotides*.

There are four different DNA nucleotides (Figure 36) that differ according to whether they contain adenine (A), guanine (G), cytosine (C), or thymine (T), and so a DNA sequence looks like an endless, apparently random string:

...AACGATCCCGAGAATGACACGGTA...

These four-letter strings look almost as boring and meaningless as the strings of 0s and 1s that make up our electronic communication, and, like those 0s and 1s, the DNA strands are

36. **DNA bases. In the four nucleotide molecules shown, the purine bases adenine and guanine and the pyrimidine bases thymine and cytosine are each linked to a 5-carbon sugar. Compared to 5-carbon ribose in Figure 33, this sugar is minus an oxygen atom. Hence deoxyribose and DNA not RNA. The ringed P is biochemists' shorthand for phosphate.**

actually packed with information. The DNA sequences in fact encode two kinds of molecule. The first kind is faithful new copies of the DNA, making it possible to pass on the genetic message to new generations and daughter cells. This depends on the structural pattern that Watson and Crick discovered. DNA molecules form a *double helix* in which the two strands are entwined rather like a spiral zip fastener. However, in a zip fastener all the teeth are identical, whereas in DNA all the way along the helix short nucleotides on one strand match long ones on the other strand and vice versa. G (long) links to C (short) and A (long) links to T (short). Thus a strand...GGATCACGTT...would have a matching strand with the sequence...CCTAGTGCAA.... When it is time to replicate the DNA, turning one double helix into two, the strands unwind and enzymes, *DNA polymerases*, read along either strand and build two new complementary strands faithfully following the rule, G to match C and A to match T.

This, however, is only an occasional task for the DNA molecules and their accompanying enzymes and would be entirely pointless without the constant 'day job'. DNA also encodes the synthesis of a different kind of nucleic acid, *ribonucleic acid* (RNA). This is done by *RNA polymerases*, reading their way along the DNA strand, again on a 1 for 1 basis, with C, G, T, and A in the DNA specifying G, C, A, and U (uridine) respectively in RNA. (Note the replacement of T in DNA by U in RNA.) Some of the new RNA molecules (ribosomal RNA and transfer RNA), as we shall see shortly, are part of the 'machinery', but the majority are so-called *messenger RNA* (mRNA) molecules with a new, distinctive coding task: mRNA molecules dictate the amino acid sequences of proteins. Clearly, with only four kinds of nucleotides and twenty different amino acids to specify, 1 for 1 coding will not work. Even with pairs of DNA letters (AC, GC, TA, etc.) one could only, in theory, encode $4 \times 4 = 16$ amino acids. It was proposed, therefore, that the *genetic code* must be a triplet code, with three-letter *codons* (GCC, AAA, CAG, etc.) specifying individual amino acids. This proved to be the case. In clever experiments, artificial, chemically synthesized RNAs with simple, repeating sequences were supplied to the protein-synthetic machinery of the cell. By examining the amino acid sequence of the artificial polypeptides thus made, it was possible gradually to work out the entire code. All sixty-four possible triplet codons are in fact used. Four are used for punctuation: three of the four, UAG, UAA, and UGA, are stop codons indicating the end of a gene. The fourth punctuation codon, AUG, does double duty, with the first AUG in a gene signalling the beginning, and all internal AUGs encoding the amino acid methionine. The other sixty codons all specify individual amino acids, and thus most of the twenty amino acids have at least two alternative codons and some three, four, or even six.

How does this work in practice? How does the sequence of codons in a long mRNA molecule get translated into the sequence of amino acids in a protein molecule? First of all messenger RNA molecules, once made, find their way to a *ribosome*. Ribosomes are made up of RNA (see above) and protein, and are the cell's assembly lines for making new protein molecules. They contain several catalytic activities of their own, so that, as the mRNA molecule gradually feeds through, and as each codon is presented, the corresponding amino acid, according to the genetic code, is added to a growing peptide chain. The key question, however, is *how* the message in RNA code is translated into an amino acid sequence. There needs to be some kind of adapter mechanism and this provides the focus for our next enzyme example.

The molecular wizardry that brings this about involves the *transfer RNA* (tRNA) molecules mentioned above and their individual enzyme handlers. For each codon in the mRNA there is a tRNA molecule with an *anticodon* to match and recognize it (Figure 37).

At the heart of the decoding process is a set of enzymes, the *amino acyl tRNA synthetases*. Each of these remarkable enzymes is highly specific for one of the twenty amino acids, but, as well as recognizing its amino acid substrate with extremely high fidelity, it has to recognize and grab the correct corresponding tRNA molecule out of the multiple set of these. Its catalytic task is to join the amino acid's carboxyl group to the end of the tRNA molecule,

37. tRNA and its anticodon loop. This shows the base sequence of the yeast tRNA for alanine. It is a flattened out two-dimensional (2-D) representation of a 3-D molecule. It is one long strand but the sequence allows four stretches of internal base pairing, giving the characteristic tRNA clover-leaf pattern. A, C, G, and U are the standard RNA nucleotides. At the X positions there is a non-standard nucleotide. Key features to note are the CCA tail, the point of attachment for the amino acid in tRNA molecules, and the anticodon loop with the sequence C-G-I (I is inosine), reading 3' to 5', to base pair with and recognize Ala codons GCU, GCC, GCA, and GCG.

making an amino acyl tRNA molecule. However, this is one of the chemical tasks mentioned earlier that will not happen without an energy drive from ATP, and so (remember we must have a common intermediate to link the splitting of ATP to the creation of an amino acyl tRNA), the first half of this reaction takes the amino acid and the ATP and creates amino acyl adenylate, splitting off two of ATP's three phosphates in the process (Figure 38).

Now the same enzyme can welcome the tRNA and, in an energetically feasible second half-reaction, transfer the amino acyl group from the adenylate onto the tRNA, making an amino acyl tRNA. A corresponding process is carried out for each of the twenty amino acids by its own amino acyl tRNA synthetase(s). The amino acyl tRNAs are now primed, ready to deliver their amino acid to the ribosome when the right codon appears. The partnership between a set of uniquely committed tRNA molecules and their uniquely committed amino acyl tRNA synthetases provides the machinery that enables the genetic code to operate, making all life possible, and the very strict substrate specificity of this remarkable set of enzymes ensures that amino acids are inserted into growing protein chains with an exceedingly low error rate.

38. **ATP in formation of amino acyl tRNA. In Step 2 the amino acid exchanges attachment to the single phosphate of adenosine monophosphate (AMP) for attachment to the terminal adenosine of the cognate tRNA.**

Isoenzymes

During the late 1950s and the 1960s, when few amino acid sequences were known and gene cloning and DNA sequencing had yet to be imagined, controversy started to emerge over a number of different enzymes' properties. These were driven by two types of experiment carried out on a particular enzymatic reaction using different organs or tissues from the same animal or plant. First of all, when measurements were made of the kinetic properties (e.g. K_m—see Chapter 2), surprising differences were observed for supposedly the same enzyme from different tissues. Second, an increasingly popular technique was *electrophoresis*, in which different proteins are separated by virtue of their different rates of movement in an electric field. Often an enzyme activity would move, not as a single band, but as several bands. In animal tissues, for example, lactate dehydrogenase frequently gave five distinct bands (Figure 39). Initially enzymologists accused one another of creating spurious artefacts through careless handling

	Heart	Kidney	Red Blood Cell	Brain	Leukocyte	Muscle	Liver
H_4	▬	▬	▬	▬	▬	—	—
H_3M	▬	▬	▬	▬	▬	—	—
H_2M_2	—	▬	—	—	—	▬	—
HM_3	—	—	—	—	▬	—	—
M_4	—	—	—	—	—	▬	▬

39. **Isoenzyme bands. Most lactate dehydrogenase (LDH) activity in humans and other mammals is encoded by one or other of two genes, one for the H ('heart') subunit and the other for the M ('muscle') subunit. LDH molecules are made up of four similar subunits and there are five possible combinations, H_4, H_3M, H_2M_2, HM_3, and M_4. H and M subunits have different charge and so the five isoenzymes can be separated by electrophoresis. Different tissues have different proportions of M and H and the patterns can be seen even in crude tissues, biopsies, etc., by staining for activity.**

of delicate enzymes. Only gradually did it become accepted that the multiple forms are a real biological phenomenon and not the result of bad experiments. Today, we have, not only amino acid sequences, but DNA sequences too, and it is clear that for many enzymes there are two or more genes, encoding distinct proteins that catalyse the same reaction. These are known as *isoenzymes* (sometimes abbreviated to 'isozymes'). Why should this be? The answer goes back to the first observation, namely the difference in kinetic properties. Higher organisms have separate organs and tissues in order to do different jobs, and so they may be using the same reaction under quite different conditions and perhaps in opposite directions. A striking example is the first reaction in glycolysis. As mentioned earlier, this is a reaction with ATP to give glucose 6-phosphate and ADP. The corresponding enzyme in muscle, studied for many decades, is *hexokinase*. Its K_m for glucose is roughly 10 micromolar (i.e. 10^{-6}M). The same activity in liver is now attributed to an isoenzyme named *glucokinase*, and its K_m for glucose is about 10 millimolar (10^{-3}M), 1,000 times higher than the value for hexokinase. The obvious thought that glucokinase is a much poorer enzyme is wrong. The liver has to process the tide of glucose and other products of digestion after a meal and thus periodically faces high glucose concentrations and needs to be able to respond over that range rather than going flat out under all likely conditions. Hexokinase, on the other hand, operates in tissues that take their glucose in a regulated fashion from the tightly controlled level in the blood, and therefore will encounter far lower glucose concentrations. Similarly, the aldolase reaction (Figure 36) is used in different tissues either to make glucose or to break it down and accordingly the tissues have different aldolase isoenzymes carefully tuned for their individual physiological tasks.

Chapter 6
Metabolic pathways and enzyme evolution

Proteins and evolution

Although a few enzymes (e.g. carbonic anhydrase) catalyse a single isolated reaction, most are part of a team that catalyses a series of reactions in which each enzyme picks up its predecessor's product, taking it a step further to create a metabolic pathway. This pathway may be to build up, say, an amino acid from simpler starting molecules, or conversely to break down food molecules to yield new chemical building blocks and sometimes also to trap useable energy. Life is the combined outcome of this seemingly logical enzyme teamwork. Like most things in the living world, this gives the appearance of purposeful planning down to the last detail. Such meticulous perfection would in past eras have been confidently attributed to the attentive skill of an all-powerful Creator. Since Charles Darwin, however, we have an alternative way of explaining how things in the living world come to be the way they are. Darwin led us to understand that natural selection could bring about stepwise beneficial adaptation over thousands or even millions of years, and, in the 150 years since the *Origin of Species*, we have learnt far more about the genetic mechanisms that can bring about such change. Does this kind of thinking work at the molecular level when we come to look at metabolic pathways and individual enzymes?

In fact the study of enzymes and other proteins allows us to be a great deal more certain than Victorian biologists could be. Many of the distinctive biological characteristics studied in comparing animals and plants, like eye colour or wing shape, have turned out to be controlled by multiple genes, whereas, in looking at individual proteins, we are looking at the products of individual genes, and latterly we can even examine those genes directly. The possibility of determining protein amino acid sequences, and, more recently, the corresponding DNA sequences, allows comparison of the same enzyme from many species and also of enzymes catalysing different but similar reactions from a single species.

Comparison across biological species

When, in the early 1960s, a trickle of amino acid sequences (only a trickle because sequencing a large protein was then still so tedious and time-consuming) started to emerge for the same enzyme purified from different organisms, biochemists were unsure what to expect. However, even these limited numbers of comparisons very quickly revealed a pattern subsequently repeated over and over again for hundreds of different enzymes and other proteins. Comparing the amino acid sequences of the same enzyme from two reasonably closely related species, say a horse and a mouse (both mammals) or a penguin and a pigeon (both birds), usually well over 80 per cent of the sequence of one aligns perfectly with the other, with only a sprinkling of mismatches, as seen in Figure 40, a horse–mouse sequence alignment for alcohol dehydrogenase with 84.5 per cent identity.

If we compare horse and frog sequences, the difference is considerably greater (68 per cent identical amino acids in the alignment for alcohol dehydrogenase), but, even more ambitiously, comparing sequences for a horse and for yeast (Figure 41), there is still discernible sequence similarity (now only 22 per cent identity). At this level, however, two other interesting

```
SP|P00327|ADH1E_HORSE   MSTAGKVIKCKAAVLWEEKKPFSIEEVEVAPPKAHEVRIKMVATGICRSDDHVVSGTLVT   60
SP|P00329|ADH1_MOUSE    MSTAGKVIKCKAAVLWELHKPFTIEDIEVAPPKAHEVRIKMVATGVCRSDDHVVSGTLVT   60
                        ************** :***:**:**.::*:*************** ************

SP|P00327|ADH1E_HORSE   PLPVIAGHEAAGIVESIGEGVTTVRPGDKVIPLFTPQCGKCRVCKHPEGNFCLKNDLSMP  120
SP|P00329|ADH1_MOUSE    PLPAVLGHEGAGIVESVGEGVTCVKPGDKVIPLFSPQCGECRICKHPESNFCSRSDLLMP  120
                        ***.: .***.*****:***** *:*********:****:**:****:.*** .:.** **

SP|P00327|ADH1E_HORSE   RGTMQDGTSRFTCRGKPIHHFLGTSTFSQYTVVDEISVAKIDAASPLEKVCLIGCGFSTG  180
SP|P00329|ADH1_MOUSE    RGTLREGTSRFSCKGKQIHNFISTSTFSQYTVVDDIAVAKIDGASPLDKVCLIGCGFSTG  180
                        ***::*****:*:**.* **:*:.**********:*:*****.****:*************

SP|P00327|ADH1E_HORSE   YGSAVKVAKVTQGSTCAVFGLGGVGLSVIMGCKAAGAARIIGVDINKDKFAKAKEVGATE  240
SP|P00329|ADH1_MOUSE    YGSAVKVAKVTPGSTCAVFGLGGVGLSVIIGCKAAGAARIIAVDINKDKFAKAKELGATE  240
                        *********** *****************:***********.*************:.****

SP|P00327|ADH1E_HORSE   CVNPQDYKKPIQEVLTEMSNGGVDFSFEVIGRLDTMVTALSCCQEAYGVSVIVGVPPDSQ  300
SP|P00329|ADH1_MOUSE    CINPQDYSKPIQEVLQEMTDGGVDFSFEVIGRLDTMTSALLSCHAACGVSVVVGVPPNAQ  300
                        *:*****.*******.**:.****************.:**  **:.* ****:****:.*

SP|P00327|ADH1E_HORSE   NLSMNPMLLLSGRTWKGAIFGGFKSKDSVPKLVADFMAKKFALDPLITHVLPFEKINEGF  360
SP|P00329|ADH1_MOUSE    NLSMNPMLLLGRTWKGAIFGGFKSKDSVPKLVADFMAKKFPLDPLITHVLPFEKINEAF  360
                        **********.*******************************.****************.*

SP|P00327|ADH1E_HORSE   DLLRSGESIRTILTF  375
SP|P00329|ADH1_MOUSE    DLLRSGKSIRTVLTF  375
                        ******:****:***
```

40. Horse–mouse alcohol dehydrogenase sequence alignment.

features assume importance. Differences are usually not randomly distributed; in some segments of the sequence the similarity is much greater than elsewhere. Also, examining the sequence differences closely, we see that at some positions evolution appears to have limited the acceptable choices. For example, if at a particular position the only amino acids found are either glutamic acid (Glu) or aspartic acid (Asp) (see Figure 19), it is clear that natural selection requires a negative charge at that position. If, as often happens, leucine (Leu) is only ever replaced by isoleucine (Ile) or valine (Val), Nature is insisting on a hydrophobic sidechain. This tendency for *conservative substitution* is evident in the stretch of sequence compared in Figure 40. Moreover, in the horse–frog alignment, 68 per cent identities imply 32 per cent non-matching pairs, 119 out of 375. If one examines these, out of 119 pairs, 83 or 70 per cent, though not identical, are similar amino acids.

The most obvious interpretation of these graded similarities is that such sequences have diverged over millions of years from a common ancestral gene, and have done so by the gradual, stepwise accumulation of mutations. Each amino acid difference is the result of mutation at the corresponding position in the gene.

Another feature, as sequences diverge, is that at a few points there may be an insertion of one or two more amino acids or a deletion, and the ends of the aligned sequences may not cleanly line up. The yeast alcohol dehydrogenase in the comparison in Figure 41 is twenty-seven amino acids shorter than the animal enzymes, which are all 374 or 375 amino acids long. Optimizing the alignments requires multiple small gaps, and one long one. If the similarity between two sequences is not immediately convincing, having to add gaps and slide sequences sideways repeatedly to display that similarity seems increasingly like cheating. However, when a 3-D structure for the protein becomes available, it usually transpires that such insertions/deletions occur on the external surface of the protein in floppy loops where the change will have little or no impact on the packing of the folded structure. Insertions or

```
P00327  ADH1E_HORSE    1   MSTAGKVIKCKAAVLWEEKKPFSIEEVEVAPPKAHEVRIKMVATGICRSDDHHVVSGTLV-
P00330  ADHI_YEAST     1   MS---IPETQKGVIFYESHGKLEYKDIPVPKPKANELLINVKYSGVCHTDLHAWHGDWPL
                          **  . * ..::*:* .  :: ::  *   ***::** *  . **

P00327  ADH1E_HORSE   60   -TPLPVIAGHEAAGIVESIGEGVTTVRPGDKV-IPLFTPQCGKCRVCKHPEGNFCLKNDL
P00330  ADHI_YEAST    58   PVKLPLVGGHEGAGVVVGMGENVKGWKIGDYAGIKWLNGSCMACEYCELGNESNCPHADL
                          *.::. *** . **:* .: ** .:  * : .  . *   *   * : : *:*  :  **

P00327  ADH1E_HORSE  118   SMPRGTMQDGTSRFTCRGKPIHHFLGTSTFSQYTVVDEISVAKIDAASPLEKVCLIGCGF
P00330  ADHI_YEAST   118   S-----------GYTHDGSFQQYATADAVQAAHIPQGTDLAQVAPILCAG
                                    :  :: .*:*  *. * *  .  * *:* . :*.  *.

P00327  ADH1E_HORSE  178   STGYGSAVKVAKVTQGSTCAVFGL-GGVGLSVIMGCKAAGAARIIGVDINKDFKAKAKEV
P00330  ADHI_YEAST   157   ITVYKA-LKSANLMAGHWVAISGAAGGLGSLAV-QYAKAMGYRVLGIDGGEGKEELFRSI
                          *  * :  .* *: * .  :  :  *** *:.  . **  * :* :* :.*::  ::

P00327  ADH1E_HORSE  237   GATECVNPQDYKKPIQEVLTEMSNGGVDFSFEVIGRLDTMVTALSCCQEAYGVSVIVGVP
P00330  ADHI_YEAST   215   GGEVFIDFTKEKDIVGAVLK-ATDGGAHGVINV-SVSEAAIEASTRYVRANGTTVLVGMP
                          *.   :*:: : :   *:*   *.:**.* .:  **:*: ***::*   .:*::.* *

P00327  ADH1E_HORSE  297   PDSQNLSMNPMLLLSGRTWKGAIPGGFKSKDSVPKLVADFMAKKFALDPLITHVLPFEKI
P00330  ADHI_YEAST   273   AGAKCCSDVFNQVVKS----ISIVGSYVGNRADTREALDFFARGLVKSPIKV--VGLSTL
                          .  :  *   :.  ::    :   * ::.*: *:* ***:*:.*:**.* :   :.:::

P00327  ADH1E_HORSE  357   NEGFDLLRSGESIRTILTF---
P00330  ADHI_YEAST   327   PEIYEKMEKGQIVGRYVVDTSK
                          *  :: :..*:*: .  :.
```

41. Sequence comparison of alcohol dehydrogenases of horse and baker's yeast.

89

deletions are much less likely to be tolerated in the tightly packed interior of a protein molecule (Box 6).

What emerges, once there are enough comparisons for a given enzyme, is a pattern that closely matches our understanding of the divergent evolution of species over hundreds of millions of years—that is, we can build gradually branching family trees for

Box 6 Scoring sequence comparisons

In comparing very closely related sequences, scoring appears to be simple and obvious. If the alignment is perfect (no gaps) and over 90 per cent of the amino acids are identical, it hardly seems necessary to do more than announce 94 per cent or 97 per cent identity. However, as discussed above, more distantly related sequences may have a majority of non-identical positions and also gaps necessary to optimize the alignment, with possibly some uncertainty as to exactly where to place the gaps. This makes it important to look more closely at the non-identical positions, especially when one is trying to decide whether sequence A is more similar to sequence B or to sequence C. There are several ways of scoring these comparisons. First of all it is important to attach a penalty to each gap, because ultimately, if you allow unlimited gaps of unrestricted length, in theory you could align any sequence with any other. When you have decided on the best alignment of two sequences you have to decide how to score non-identical amino acids. As mentioned above, there is a strong tendency for amino acids at important positions to be replaced at the very least by similar amino acids. Accordingly there are a variety of scoring tables available based on similarity. This, however, is highly subjective: should one emphasize charge, hydrophobic character, shape, size, etc.? Also one can see that the scores ought to be different for small proteins (large surface to volume ratio) from those appropriate for large proteins, and again for proteins in free solution compared to proteins associated with membranes. In the early days, when the comparisons were all done for the proteins, rather than the genes encoding those proteins, there were two serious attempts at less subjective scoring. One was based on 'minimum mutational distance', looking at the possible codons for each amino acid and working out the smallest number of single-base changes in the DNA that could account for the amino acid change. This, however, is bound to be a serious under-estimation of the real mutational distance. Another was a table of 'accepted point mutations'. This was based on compiling all known sequences for proteins and working out, without any subjective judgements, how many times, in fact, methionine is replaced by leucine, glutamic acid, etc., so that one ends up with a table giving each amino acid an empirical score for each of the nineteen possible replacements.

Nowadays, there are vastly more sequences available, and it is very seldom necessary to guess at the likely DNA sequence, because overwhelmingly now sequences in the databases are likely to have been determined for the gene. In other words the protein sequence is likely to be worked out from the DNA sequence rather than vice versa. With so many sequences to analyse it has become important to have better tools and there are now powerful algorithms for carrying out *multiple sequence alignment*. The most widely used of these is CLUSTAL, devised by the Irish biologist Desmond Higgins. It remains important, however, to think about the purposes of the comparison in deciding on scoring procedures for gaps and similarity.

each enzyme. We can also see that some changes have clearly been selected because they offer an advantage in their context, whereas others appear to have been tolerated because they cause no harm. It is easy to see that such benign mutations could become fixed in a small isolated population. Now that DNA sequencing has been applied on a vast scale to human populations for some twenty years, it has become clear that our own genome is scattered with such 'random drift' mutations, and their geographical distribution allows us to draw many conclusions about the movement of ancient populations.

Enzyme families

Taking a step further in exploring molecular evolution, we meet the idea of *enzyme families*. Thus far we have considered the similarity in different biological species (say, a cat, a cauliflower, and a centipede) between different versions of what is fundamentally the same enzyme doing the same chemical job. However, we can also search for similarities between enzymes doing different jobs. It turns out frequently that, if the jobs, though different, are chemically similar, the enzymes may display sequence similarities. Two enzymes well known to biochemists are *lactate dehydrogenase* and *malate dehydrogenase*. The shared 'dehydrogenase' tag tells us that they catalyse similar chemical reactions (oxidations). Both use the same cofactor, NAD^+, and both work on hydroxyacids, lactic acid and malic acid (which exist, at physiological pH, in their charged forms, the lactate and malate ions). However, these are different compounds (Figure 42), and accordingly lactate dehydrogenase will not work with malate and malate dehydrogenase will not work with lactate. Nevertheless their amino acid sequences show clear similarity (Figure 43) which can also be seen in the 3-D folds of their molecules.

Divergence and convergence

The story so far relates to a model in which similarity reflects divergence from a common ancestor, whether recent or distant. However, biology has already taught us that evolution can play tricks. If a particular characteristic is beneficial, say body shape for running fast, or wing shape for silent flight, unrelated species might arrive at the same destination by *convergent* evolution.

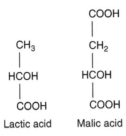

COOH
|
CH₃   CH₂
| |
HCOH   HCOH
| |
COOH   COOH

Lactic acid   Malic acid

42. Lactic acid and malic acid.

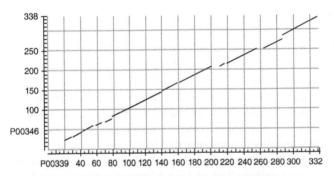

43. Sequence alignment of lactate dehydrogenase and malate dehydrogenase. The two sequences run along the two axes and the algorithm compares every position in one with every position in the other, assigning a score. A window size, for example, ten or fifteen residues, is assigned and moved along each possible alignment. When a consistently high score is obtained for an alignment as the window moves along, it generates a diagonal line.

Often, similar ecological niches in different continents may be occupied by species that look very similar but are in fact only very distantly related. Is this also the case for enzymes? Can we be confident that the similarity at the molecular level between lactate dehydrogenase and malate dehydrogenase unequivocally indicates divergence from a common ancestral enzyme or might it instead indicate convergence on a pattern that works well?

In fact, enzymes offer striking examples of both divergent and convergent evolution, and it seems they can be readily distinguished. In the case of individual enzyme families, one of the best-documented examples of both divergent and convergent evolution comes from the proteolytic enzymes. In Chapter 5 we met trypsin, chymotrypsin, and elastase. All of them catalyse breakage of peptide bonds, with different preference in each case for favourite amino acids forming the peptide bond. For each of these enzymes we have multiple sequences to compare and we also have a number of 3-D structures. From these, and from much study of their functional properties, we know that these enzymes form a 'family'. This means not just that all the trypsins are related by divergent evolution but also that all the trypsins are related to the chymotrypsins and the elastases. We can see them as different models of the same fundamental molecular machine, rather like related car models doing slightly different jobs. They are all categorized as *serine proteinases*. This is because they all have one particular residue of the amino acid serine in their structure that plays a key chemical role in their catalysis. This is what we term an *essential residue*, meaning that, if that serine is either genetically replaced or chemically altered or blocked, the enzyme is almost catalytically dead. In each enzyme's active site, assisting the serine in its task, are two other amino acids, an *aspartic acid* residue and a *histidine* residue. Together these three make up what proteinase lovers call the *catalytic triad*. By comparing the 3-D structures one can see that the geometrical arrangement of these three amino acid residues in different members of this serine proteinase family is essentially identical. One can also see the key differences that distinguish the different members, allowing each to exercise its particular preferences for its favourite kind of peptide bond in a protein substrate.

Serine proteinases similar to chymotrypsin are widely distributed in Nature and are found even in bacteria. However, bacteria also contribute another group of proteinases called *subtilisins*. It turns out that these too are serine proteinases and use the same chemistry as chymotrypsin and its cousins. However, lining up the amino acid sequence of (any) subtilisin with that of (any) chymotrypsin, trypsin or elastase, reveals no significant similarity (i.e. no more matches than the chance matches one gets from lining up any protein amino acid sequence with any other unrelated sequence). This is a second serine proteinase family unrelated to the first. Not surprisingly then, X-ray crystallography reveals that members of the subtilisin family have a quite different folded shape from that of chymotrypsin, etc. However, the 3-D structures also allow us to see that the three members of the catalytic triad in subtilisins are the same three amino acids in precisely the same geometry in 3-D space as in the chymotrypsin family. This is in spite of the fact that the three triad members are not even in the same linear relationship in the amino acid sequences of the two families (Figure 44). These two families thus provide a remarkable example of convergent evolution at the molecular level, but show that, unlike a bird's wing, two proteins do not need to have the same overall shape or a similar sequence in order to offer the same chemistry. Identical catalytic machinery can hang off entirely different, unrelated protein scaffolds.

So it seems that, where we find convincing sequence similarity, we can safely assume it reflects divergent evolution. Indeed it is by also resorting to the molecular level of comparison that we can be

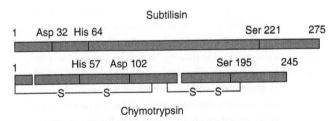

44. Catalytic triads in chymotrypsin and subtilisin. Note the completely different position, order, and spacing of the three catalytic triad amino acids. In chymotrypsin, note also the disulfide bridges and the two gaps introduced in activation of chymotrypsinogen.

so confident nowadays of which species are examples of convergent evolution. If, say, the colour and size and wing and beak shape of an American bird and an Australian bird are very similar but their DNA or protein sequences do not show close similarity, we are left in no doubt.

Where do new enzymes come from?

In our discussion of enzyme families there is an issue we have not yet addressed. Assuming, for the sake of argument, that contemporary malate dehydrogenases are descended from an ancestral lactate dehydrogenase (or perhaps vice versa), how do we face the fact that creatures like ourselves possess and require *both* these enzymes? It is clearly not sufficient to postulate one gradually morphing into the other over evolutionary time. An answer to this apparent puzzle comes from experiments with bacteria. Microbiologists can grow bacteria in *chemostats*, vessels in which the experimenter controls all the conditions and sets up a steady inflow of nutrients matched by an equal outflow, so that the volume stays constant. The bacterial cells grow and multiply and reach a limiting population density that is dictated by the nutrients. That population density is measured in the outflow and will remain constant over very many generations of cell division so long as the conditions remain the same. The experimenter can then impose a challenge, for example by severely restricting the supply of one of the key nutrients. If so, the population density will at first drop sharply to a new, much lower constant level, but if it is now monitored over many days, often there is a sudden recovery in the population density. When this happens it is the result of a beneficial mutation. This is natural selection in action in the lab. A mutant cell with an advantage will multiply and progressively take over the entire population. There are various ways in which a mutation could improve the life prospects of an individual bacterium, but one way that is crucial to our puzzle is to make an entirely new copy of an existing gene, so that the mutant cell now has two copies and can make double the amount of the corresponding enzyme. This process of *gene duplication* is readily demonstrated and occurs quite often.

This response can, of course, be adopted by cells in the wild in answer to an environmental stress. What happens now if the stress is removed and good times return? The extra copy of the gene is now redundant and over time it is likely to be eliminated. On the other hand, over that period of time, one of the bacteria carrying the two gene copies might pick up a random mutation in one of those copies. Occasionally such a mutation might confer new functional properties. Specifically, it might allow the encoded enzyme to work with a new substrate. Meanwhile the other gene copy continues to cater satisfactorily for the original requirement. The new enzyme might, of course, be entirely useless, in which case the gene would still be eliminated over time. However, there is also the possibility that the new activity might open up new possibilities for the organism, in which case it will be retained and will come under separate selective pressure to further improve.

This sequence of events provides a way for evolution to create new enzymes without losing the old ones. It explains how enzyme families like the hydroxyacid dehydrogenases or chymotrypsin and related serine proteinases arise.

How did metabolic pathways arise?

Returning to the questions posed at the beginning of the chapter, explaining how enzyme families can arise still does not account for metabolic pathways. Most accounts of the early stages of life on earth envisage it arising in an environment where there was little or no free oxygen and where conditions were favourable for some of the main building blocks of life to arise through spontaneous chemistry. Many attempts have been made to mimic the likely conditions and show that amino acids and the building blocks of DNA and RNA do arise without biological activity (other than that of the experimenter). This leads to the concept of the 'primaeval soup'.

According to these ideas, early life forms might have had it easy, with a varied menu of ready-made compounds on offer, and therefore only a limited need for complex metabolic pathways to make vital compounds. This led to a theory of the evolution of metabolic pathways advanced by Norman Horowitz in 1945, at a time when the nature of genes was not yet known and the 1:1 relationship between genes and enzymes was just being established. Horowitz postulated that biosynthetic pathways must have evolved backwards: as a key compound, Z, in the primaeval soup was used up it would be necessary to evolve a catalyst to make it from a related compound, Y, that was still available in large amounts. In time, Y too would start to run out and so there would be pressure to evolve a new enzyme to make Y from X and so on. A key thought in this hypothesis of *retroevolution* was that the enzyme to convert Y into Z and the enzyme to convert X into Y both have to recognize compound Y and so, perhaps, with suitable evolutionary tweaking, one could be the precursor of the other. All the enzymes in a pathway would then be related, with closest similarity between enzymes catalysing consecutive reactions. Seventy-five years later, with the benefit of far more knowledge of how enzymes work, we can see that the first part of the hypothesis is likely to be correct, but the idea of each enzyme in a pathway being directly related to the next is not persuasive because the nature of the chemistry at each step is different from the last or the next. We can now see that new enzymes are likely to be recruited, not along pathways, but across pathways as in the case of lactate dehydrogenase and malate dehydrogenase, which belong to different pathways. Evolution usually finds it easier to tweak substrate specificity than to radically alter the nature of the chemistry. In recent decades, humankind has unintentionally carried out an experiment that illustrates how this can happen. Through our widespread over-use of antibiotics, we have thrown down the gauntlet to bacteria, which have responded by effectively searching their existing repertoire of catalysts for one that can most readily be adapted to tackling the new molecular challenge (see Chapter 7).

Still further back

This makes a satisfyingly plausible account of an important stage in biomolecular evolution, but it still leaves a major puzzle further back in relation to the origins of life. We cannot produce proteins, and therefore enzymes, without the DNA instructions that encode their amino acid sequences. That suggests that nucleic acids must be the 'chicken' and proteins the 'egg'. But, on the other hand, if we consider the production of DNA, its transcription to form an RNA 'message', and the translation of that message to form proteins, all these processes are carried out by highly specific enzymes that perform their task with incredible fidelity (very low error rate), suggesting the opposite hierarchy. Which came first? Each requires the other and for a long time this seemed an insoluble puzzle, with proteins able to offer potent catalysis but not flexible storage of information. Nucleic acids with their four-letter genetic code are ideal for storing information and, critically important, for templating their own replication. But, seemingly, nucleic acids did not offer the structural diversity to provide versatile catalysis. However, detailed study of ribosomes, the subcellular particles that carry out protein synthesis, has changed that. They are made of both protein and RNA, and this RNA contains not just the standard four G, C, A, and U building blocks but a number of others, and it is now clear that some of these RNA molecules do indeed carry out catalysis, leading to the concept of *ribozymes*. In 1989, two American scientists, Thomas Cech and Sidney Altman, received the Nobel Prize for Chemistry for their separate discovery of RNA catalysis. We can no longer state that all enzymes are proteins. This still leads on to many new questions but nevertheless at last it offers a plausible partial solution to the puzzle and a new vision of an early 'RNA world' in the early stages of the emergence of life on our planet. It starts to look as if enzymes are, after all, the egg and not the chicken.

Chapter 7
Enzymes and disease

Enzymes in a medical context

In relation to medical science, we need to think of enzymes in two quite different ways. They may be the problem or they may offer the solution. In the first context, what happens if enzymes are faulty in some way? As we have seen, the smooth running of our bodies relies on many hundreds of different enzymes. Each one is a complex piece of molecular machinery, and, like ordinary machinery, each one can go wrong in a variety of ways, leading to more or less serious clinical symptoms. On the other hand, there are often situations in which we deliberately seek to damp down the activity of normally functioning enzymes in our bodies, and this is how many drugs work. Enzymes, human or otherwise, are also nowadays widely used as agents for diagnosis or therapy.

Enzymes for diagnosis

Since we are able to isolate single enzymes in a pure state, we can use them as diagnostic tools. When a doctor sends off blood or urine samples for 'tests', often the chemical pathology lab will use enzymes for accurate and specific measurements of tell-tale compounds such as cholesterol, urea, glucose, etc. Equally, the robust devices used in the home or on the road for self-monitoring glucose in diabetics or measuring alcohol in a driver's breath rely on purified enzymes.

Enzymes can be used in another way for diagnosis. Different organs in our bodies have distinctive complements of enzymes. If there is tissue damage enzymes spill out of the broken cells into the bloodstream. Signature enzymes can therefore give early warning of liver or kidney damage or a heart attack. This needs a specific measurement of the presence or absence of the indicator enzyme in a blood sample and of the level of activity if present.

Sick enzymes

In the first decade of the 20th century an English physician, Archibald Garrod, became interested in rare diseases that appeared to run in families. In one of these diseases, *alkaptonuria*, the urine of newborn babies turned black. Garrod showed that the urine contained large amounts of homogentisic acid (Figure 45) which reacted with oxygen in the air to form the coloured product. Normal babies also make homogentisic acid as one of several steps in the breakdown of the amino acid phenylalanine arising from food proteins. However, in alkaptonuria, the enzyme activity that catalyses the onward reaction of homogentisic acid is missing, and so, like water behind a dam, this compound accumulates. For this and similar diseases, Garrod coined the term '*inborn errors of metabolism*'. He rightly deduced that these are genetic diseases and that they obey *recessive inheritance*: only if you inherit a 'bad' gene from both parents are you affected. (Accordingly such diseases are far more common in communities that encourage first-cousin marriages.) He also postulated that the genetic trait was the presence or absence of an enzyme activity, a remarkable insight, given the very limited understanding of both enzymes and genes at that time.

The pattern in alkaptonuria is typical. A blockage in a metabolic pathway inevitably leads to a pile-up behind the blocked step. A normal metabolite is thus found in very abnormal concentrations in blood, urine, and other bodily fluids. Often the accumulated compound also spills over into other pathways and reactions,

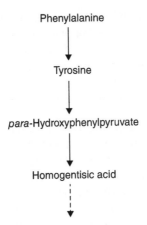

Phenylalanine

↓

Tyrosine

↓

para-Hydroxyphenylpyruvate

↓

Homogentisic acid

⋮
↓

45. Early enzymatic steps in the breakdown of phenylalanine.

leading to the appearance of entirely *abnormal metabolites*. Finally, our bodies deploy a range of detoxification reactions that are normally used to handle foreign compounds, making them less harmful and easier to excrete. Each of these diseases therefore gives rise to its own distinctive pattern of unusual metabolites that can readily be detected and measured for diagnosis.

Another disease affecting phenylalanine metabolism is PKU (*phenylketonuria*), first described by Asbjørn Følling in 1934. This is much more common than alkaptonuria, much less obvious in announcing its presence, but very drastic in its long-term effects if undiagnosed. PKU patients have a mutation in the gene for the enzyme *phenylalanine hydroxylase*, which catalyses the first step in phenylalanine breakdown (Figure 45). This insidious disease used to lead unfailingly to intellectual impairment, and in some countries a high percentage of the inmates of mental institutions turned out to be PKU patients. Restricting protein intake could lessen the extent of brain damage, but timely intervention required early diagnosis. This arrived in the 1960s thanks to Robert Guthrie, an American microbiologist motivated

103

by the intellectual disability of his own son and niece. He introduced a simple procedure to test a drop of blood taken from a newborn baby's heel for excessive levels of phenylalanine, making dietary intervention possible before there was time for significant damage. (Fortunately, up to the moment of birth, the mother's enzymes take care of the problem, preventing damage *in utero*.)

This was the beginning of routine neonatal screening, applied today in many countries, not only for PKU but also to detect a number of other genetic diseases. The range and sensitivity of measurement methods has improved steadily in recent decades. Most often these rely on detecting abnormal metabolites (Figure 46), but other approaches may measure enzyme activity directly or detect the protein itself, and latterly DNA methodology has made it possible to diagnose defects at the level of the gene, by looking for the actual mutation.

This last advance has shown up very clearly that the traditional all-or-none, active/inactive approach reflected in classical genetic family trees is an over-simplification. For a typical enzyme protein there will be several hundred positions where a mutation could change the amino acid. Many such changes could be harmless, but there are bound to be numerous positions where a change might remove a key chemical player in catalysis or perhaps interfere with the protein's ability to fold to the right active shape. So, for any disease that, like PKU, affects many thousands of people all over the world, the DNA studies show that there are many different disease mutations. Their uneven geographic distribution gives clues to the movement of human populations. The identification of the precise mutations can also have direct clinical relevance since some mutations are less severe, perhaps leaving a residual level of enzyme activity rather than abolishing it totally.

New sources of precise knowledge inevitably present new choices and tricky ethical issues. For example, some diagnostic tests can be applied not only at birth but also early in pregnancy. In some

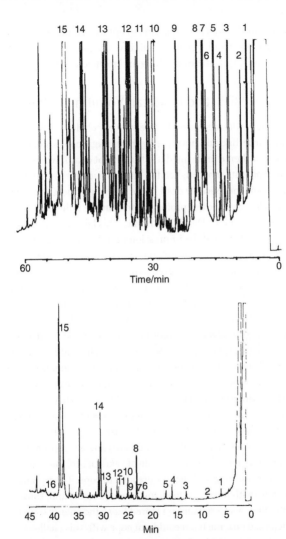

46. **GC (Gas Chromatography) for diagnosis.** Each 'peak' shows a
separate compound identified by mass spectrometry. The traces are
for urine samples from children with a gene defect in fat oxidation.
The upper trace is for a sick child on admission and shows many
unusual metabolites. The lower trace is for a similar case under
dietary management.

jurisdictions this will allow parents to opt to terminate a pregnancy. Above all, however, definitive diagnosis opens the door both to reliable genetic counselling and to effective therapy.

Enzymes, SIDS, and Jamaican Vomiting Sickness

Some enzyme defects make themselves very obvious. Others may be borderline and nevertheless have tragic consequences. A recurrent theme in the media is that of 'cot death', where parents put an apparently normal, healthy baby to bed at night only to find the baby inexplicably dead in the morning. These mysterious tragedies led the medical profession to coin the unhelpful label SIDS, Sudden Infant Death Syndrome. This label disguised total ignorance and created the impression that this was a single disease rather than a single outcome attributable to many different causes. I was involved in a medical collaboration in the 1980s that revealed that one possible cause could be a congenital enzyme defect in the handling of fat. To use fat as an energy source we break down *fatty acid* molecules that are initially sixteen or eighteen carbon atoms long and break off 2-carbon fragments, giving 14-, 12-, 10-, 8-carbon intermediates, etc. Even though this is repetitive chemistry, it is not surprising, in view of what we know about enzymes, that a single enzyme can't cope with the entire range of lengths. So we have separate enzymes to handle long, medium, and short chains. Crucially, though, there is a bit of overlap, because the 'cot death' cases in question turned out to have a defective medium-chain enzyme. For most of the time, and indeed for the first couple of years of life, the outer pair of enzymes, for long and short, were able to cope. The tipping point came with missing a feed, low blood glucose, and, thus, exceptionally heavy reliance on stored fat as the energy source. The depleted enzyme team could not cope with this challenge, leading to crisis and death.

Recognizing the genetic cause of such a cot death offers relief to guilt-ridden parents and provides a basis for genetic counselling, that is, advice to intending parents on the risks for a child as yet unborn or possibly still to be conceived. Similarly to PKU, this is a defect that, provided it is diagnosed, can be handled by dietary management—in this case (1) simply avoiding fasting and (2) maintaining a frequent, high-carbohydrate, low-fat feeding regime (see Figure 46).

The darker side of this story is that parents in such cases have often had their tragedy compounded by suspicion that they have killed their child. In one particularly sad case a few years ago an 'expert' witness informed a UK court that the chances of a couple's first cot death through natural causes were 1 in 8,000, and, that, following the second cot death in the same family, the chances of that were 1 in 8,000 x 8,000 (i.e. 1 in 64 million). The mother went to prison because the 'expert' failed to understand that, if the first death, however unlikely, had a genetic cause, then the chances of the same defect affecting the next child were 1 in 4. For a recessive condition (see above), the assumption is that the two parents are *carriers*, each carrying one 'bad' copy and one 'good' copy of the gene. They are therefore unaffected themselves but both have a 50:50 chance of passing on the bad copy to their offspring. Hence a 25 per cent chance that the child will be affected and a 50 per cent chance that the child will be unaffected but also be a carrier like the parents.

Another entirely different indication that children need their fatty acid oxidation enzymes comes from the West Indies, where popular songs celebrate salt fish and ackee fruit. The fruit, a member of the lychee family, was brought over in 1793 from West Africa by Captain Bligh (of *Mutiny on the Bounty* fame). It is unfortunately so delicious that little boys climb up in the trees to steal the fruit while it is still unripe. In this state the ackee contains high concentrations of an unusual amino acid, Hypoglycin A, so called because it drastically lowers blood sugar levels. This turns out to be an example of *lethal synthesis*: Hypoglycin A is superficially so similar in size and shape to the normal protein amino acid leucine that it is accepted by the normal breakdown pathway, going through several steps until a highly reactive metabolite is produced that irreversibly blocks one of the enzymes for short-chain fatty acid oxidation. This final step is a case of *suicide inhibition*, in which the enzyme's attempt at catalysis on a dangerous substrate destroys its own activity. The clinical result for the little boys, Jamaican Vomiting Sickness, can be fatal.

Enzymes as targets

Broadly, there have been three approaches to the development of drugs. First, over most of human history, experimentation with the products of the natural world has led to accumulated wisdom about the efficacy of various poisons, potions, and drugs. Our ancestors had no knowledge of chemistry, biochemistry, or physiology, and yet over centuries they came upon many potent remedies. Only over the past century have we been to explain how and where these substances work and then, in some cases, to improve upon the original folk remedy. Second, the explosion of chemical knowledge and competence from the mid-19th century on made it possible to synthesize a huge range of new compounds. Large-scale screening of such compounds has turned out to be, even if uninspired, a remarkably effective way of discovering new drugs. In recent times the pharmaceutical industry has automated the 'high-throughput' screening of vast combinatorial 'libraries' of such compounds.

With both of these approaches, many of the effective drugs turn out to be enzyme inhibitors: compounds that attach to an enzyme molecule and block, or at least decrease, its activity. With the advance of biochemical knowledge this in turn has led to the third approach, rational drug design, in which you know the enzyme target, you know its structure, and thus in principle you should be able to design the ideal drug.

Aspirin

The quantities of aspirin sold worldwide to control pain and fever and to inhibit blood clotting amount to 40–50 tons annually. This drug is based on a traditional remedy used over thousands of years: an extract of willow bark or leaves. Chemical investigation in the 19th century established that the active ingredient was salicylic acid, a very simple organic compound. This compound is very bitter and Felix Hoffmann, at the Bayer dye company in Germany, modified it by converting it to acetylsalicylic acid, aspirin. It took another seventy years before Sir John Vane at the Royal College of Surgeons in London discovered how aspirin works. It inhibits the enzyme *cyclooxygenase*, which catalyses a key step on the route to a class of inflammatory compounds called prostaglandins. The inhibition involves chemical transfer of the acetyl group to block the enzyme's active site. It seems that Hoffmann had fortuitously modified salicylic acid in exactly the way our bodies do to create the active inhibitor.

Warfarin

During the Great Depression in the USA, farmers could not afford to discard mouldy hay as fodder. A resulting outbreak of cattle deaths from internal bleeding was traced back to *dicoumarol*, a compound present in spoilt clover. We now know that this acts as an inhibitory analogue of *Vitamin K*. This vitamin is required for a remarkable enzymatic process that re-tailors several of the cascade proteinases involved in blood clotting (Chapter 5), inserting CO_2 into particular glutamic acid sidechains. This gives these sidechains an extra carboxyl group (COO^-), forming a pair of claws to grab calcium ions which play a key role in proteinase activation (Figure 47). Research led to the development of further synthetic compounds that could control thrombin activation in the same way. One of these is *warfarin*, more generally known as rat poison. This can be used as a hostile compound to kill pests,

47. Gamma-carboxyglutamic acid. Carboxylation of glutamate sidechains in some enzyme proteins, for example blood clotting proteins, forms a dicarboxylate 'claw' very effective at holding calcium ions (Ca^{++}).

but also, in smaller doses and with care, it can be used therapeutically to avoid internal clotting in elderly patients.

Penicillin

Some of our most important drugs are *antibiotics*, intended to kill hostile invading organisms. If such a drug is an enzyme inhibitor, ideally it should inhibit an enzyme that is found in the infectious agent but not in our own cells. Penicillin is a perfect example. This compound, fortuitously discovered by Alexander Fleming in 1928, and later developed by Howard Florey and Ernst Chain, is produced by the *Penicillium* mould and it kills bacteria by sabotaging one of the enzyme reactions that they use to build their tough cell walls. Human cells do not have these tough walls and so are not affected by the drug.

This 'wonder drug' had a major effect on saving the lives of wounded soldiers in World War II and the three scientists principally involved received the Nobel Prize in 1945 for their breakthrough. Penicillin has also subsequently provided a worrying example of evolution in action. The bacteria have been able to fight back in recent decades by evolving enzymes of their own, β-*lactamases*, that attack the penicillin molecule, rendering it harmless. Similar *drug resistance* enzymes have emerged to counter the action of various more recent antibiotics, leading to a resurgence of dangerous diseases such as tuberculosis. Unfortunately for us, drug resistance too is a manifestation of the power and versatility of nature's toolbox.

Captopril

A very widely used class of drugs nowadays are the '*ACE inhibitors*'. ACE stands for angiotensin converting enzyme and carries out a conversion step (cutting a peptide bond—Chapter 5) making a peptide molecule that we use naturally to control our blood pressure. ACE inhibitors dampen this down in such a way as to dilate our blood vessels and so lower blood pressure. Captopril was the first in a line of such drugs and is generally regarded as the first clear example of a *designer drug*. The work was unwittingly assisted by Brazilian tribesmen who use the venom of pit vipers as an arrowhead poison. The venom contains a peptide that is an ACE inhibitor, and this was used as the starting point for molecular modelling to design a perfect synthetic molecule that would fit snugly into the active site of ACE and block its action.

Note that, once again, the self-same compound can be either beneficial drug or poison, depending on the dosage.

Viral infection: HIV, coronaviruses, etc.

In the early 1980s fear spread among a whole generation of young people with the emergence of a killer epidemic, a new sexually transmitted disease, AIDS. The name AIDS (Acquired Immunodeficiency Syndrome) conveys the essential fact that this disease disarms our normal immune response mechanism leaving a patient very vulnerable to other infections. This dangerous disease turned out to be the result of invasion by a virus, HIV, the human immunodeficiency virus. The crisis and ensuing stream of young deaths led to an intensive international collaboration to develop weapons against the viral invader. This was remarkably successful over just a few years, so that today to be diagnosed HIV-positive is no longer regarded as a death sentence. Some of the initial drugs were directed at inhibiting the RNA polymerase involved in the virus's nucleic acid production, but a key step in developing one of the drugs that goes into the effective treatment cocktail today was the solution in 1989 of the structure of the *HIV proteinase*. Like all viruses, in order to multiply and move on in life (by infecting new cells), HIV has to package each of the multiple new copies of its genetic material in its own purpose-built protein coat. Virus coats are geometric structures assembled by putting together identical building bricks, like molecular Lego. For HIV, a critical step in making these bricks is cutting down the protein to the right size, for which purpose, fortunately, it uses its own dedicated proteinase. Thus, solving the structure of this proteinase meant it was possible to design specific potent inhibitors. There are now a range of these compounds which very effectively stop the virus in its tracks.

More recently the world has been hit by pandemics caused first by the Ebola virus and then by a series of major outbreaks of severe respiratory diseases caused by various coronaviruses—SARS, MERS, and most recently and devastatingly SARS-CoV-2, causing the disease named COVID-19. Once again, scientists across the world have leapt into action, and it is really striking how, with the exception of vaccine development, every aspect of this medical crisis involves enzymes, either of the host victim or of the virus itself. First of all, every virus has to gain entry to its target cells and does so by having a coat protein with a high affinity for a molecule on the cell surface. In the case of coronaviruses the protein with this task is the 'spike protein' that, in multiple copies, decorates the surfaces of these viruses, giving them the appearance of a curled-up hedgehog. The human protein that is the target of the spike protein is the enzyme ACE2, forming part of the system for controlling blood pressure just mentioned above. This protein begins its active life anchored on the surface of cells. Once the spike protein docks with one of the ACE2 molecules, this enables the virus membrane to fuse with the human cell membrane, spilling its contents into the cell and launching the takeover of the machinery.

Inside the invaded cell, among the proteins encoded by the viral RNA is the RNA polymerase required to turn out multiple copies of the viral genome for new viral particles. (Note that this is an RNA virus, meaning that it stores and hands on its genetic information as RNA, not DNA.) This, being a viral enzyme, is an obvious drug target, and, since drugs had already been developed to attack the Ebola and HIV RNA polymerases, controlled trials of these drugs were rapidly launched to assess their effectiveness against SARS-CoV-2. At the time of writing, the Ebola drug Remdesivir is showing promising results. SARS-CoV-2 also has to produce and process its coat proteins. To do this it makes use of a highly selective proteinase called furin. Unlike the HIV proteinase, however, this is not a likely drug target, because it is a human enzyme used to process a number of hormones, so that inhibiting it would certainly have unwanted side-effects.

Another focus of major effort in the fight against SARS-CoV-2 has been the search for reliable test procedures. Some of these rely on specific detection of the viral RNA, utilizing enzymes for PCR (*polymerase chain reaction*—see Chapter 9) analysis. Others use antibodies to detect the viral proteins, and here a vital element is the principle of enzyme-linked antibodies. Potent and selective though it may be, a colourless antibody molecule does not offer a brash signal of its presence. If, however, one chemically links the antibody to an enzyme (horseradish peroxidase is a favourite) it can now produce a visible, measurable signal in a very short time via the catalytic reaction. (In addition to detecting viruses, this method is the basis of a large number of so-called ELISA, or Enzyme Linked Immunosorbent Assay, routine diagnostic tests—e.g. pregnancy tests.)

Therapeutic enzymes

An obvious role for therapeutic enzymes is in the treatment of the inborn errors of metabolism discussed above. As we have seen some of these diseases can be dealt with by dietary control, but it would be better to mend the defect. Since the fundamental problem is at the level of the gene, one approach is *gene therapy* Much effort is going into ways of introducing the correct gene into at least some of the patient's cells, but this is not yet entirely successful. An alternative is to supply the enzyme. This has been or is being attempted for a number of genetic diseases, including PKU. There are several tricky issues. How and where will the enzyme be delivered? By repeated injection? Will the replacement enzyme be human or not? If the enzyme is from some other convenient source it will provoke an *immune response* in the patient. In the case of one of the enzyme therapeutics on the market for PKU this is circumvented by *PEGylation* covering the foreign protein with an innocuous 'cloak' of polyethylene glycol (PEG).

The PEG trick is also used with *asparaginase*, a therapeutic enzyme used for treating certain forms of leukaemia. The cancer cells are unable to make their own asparagine, one of the twenty amino acids needed to make proteins. Asparaginase prevents these cells from simply taking up asparagine from the blood by breaking it down. PEG masking is also used to deliver *uricase* to joints affected by gout, which results from the build-up of insoluble uric acid.

The issue of immune rejection is less of a problem when the enzyme is not injected into the bloodstream. Thus, in cases of pancreatic insufficiency, capsules can be taken by mouth to deliver non-human trypsin and lipase. Similarly the enzyme *collagenase* can be successfully applied directly to skin ulcers.

A very successful therapeutic drug has been tPA, *tissue plasminogen activator*. In Chapter 5 we met plasmin, an enzyme that prevents the blood clotting mechanism running out of control. However, life threatening blood clots do form sometimes in the blood supply to the heart (coronary thrombosis) or the brain (stroke). The outcome depends very much on whether the clot can be quickly removed or cleared, and the chances have been greatly improved since it became possible to inject tPA to stimulate our own anti-clot enzyme defence.

Enzyme conjugates, ADEPT

In the management of cancer, a big problem with traditional methods has been that the patient can perceive the treatment as almost worse than the disease, since the chemicals given to attack the cancer cells also attack many other types of cell. In the frontline of cancer drug development today is a sophisticated range of drugs with names ending in 'mab', standing for monoclonal antibody. One half of the therapeutic molecule is an antibody against specific proteins on the surface of the target cancer cells and the other half is an enzyme. (This strategy relies on the surprising fact that, often, if the genes for two proteins are joined end to end, they successfully encode an enlarged protein combining both functions—i.e. both halves still fold successfully.) When this delivery vehicle reaches its destination, the whole molecule is taken up into the cells. The other half, chemically attached to the antibody, is the enzyme 'warhead' that inflicts the damage. The enzyme itself could be one that, for example, attacks the cells' nucleic acids. Alternatively, in an approach known as ADEPT (antibody-directed enzyme prodrug therapy), the enzyme releases a potent drug inside the cell from an apparently innocuous precursor, the 'prodrug'. The enzyme carboxypeptidase

(Chapter 5) has, for example, been used to release a poisonous 'nitrogen mustard' from peptide linkage to an amino acid. These approaches largely eliminate the side effects associated with traditional chemotherapy.

Enzymes have come to play a major role in modern medicine, pharmacology, and chemical pathology, sometimes as the centre of the story, sometimes as tools or as markers. As our detailed knowledge of the biochemistry of disease advances, this role can only increase.

Chapter 8
Enzymes as tools

Thinking beyond the original biological context

As we saw in earlier chapters, the first preoccupation in the study of enzymes was to work out what each one does and how their combined action achieves biological objectives. An essential stage in this journey of discovery was the isolation of individual enzymes in a pure state. In due course, many hundreds of purified enzymes found their way from the research laboratories into the catalogues of scientific suppliers. In the first instance they were immensely valuable as research tools. In studying a particular enzyme one might gratefully employ other available pure enzymes, perhaps to supply the substrate or to measure the product of a reaction. In this context, enzymes tend to be bought and sold in milligram amounts. But enzymes have now burst out of the narrow confines of research laboratories and onto the industrial stage, where they are marketed sometimes on a scale of tons.

Overcoming barriers to application

In the past there were considerable barriers to this wider application of enzymes. First of all, purification was labour-intensive so that pure enzymes were inevitably costly. The degree of purity required would depend on the application, and for some applications a relatively crude preparation might suffice. Second, however, as mentioned in Chapter 3, typically enzymes were fragile and unsuitable for robust industrial handling. With regard to purification, from the 1960s onward there were many improvements in methods, in particular in the range of materials available for chromatographic purification, so that, instead of a pathetic yield after six to eight steps, a typical purification might now give a generous yield of pure enzyme after only three to four steps. However, the biggest advance came through the advent of gene cloning in the 1970s.

This involves cutting a particular gene of interest out of the source DNA (with an enzyme—see Chapter 9) and neatly placing it in a *vector*. Most commonly the vector is a *plasmid*, a circular piece of DNA able to carry the gene into a foreign 'host', that is, another organism that will replicate the plasmid, and therefore also the cloned gene (Figure 48). This is sufficient if the objective is simply to determine the DNA sequence of the gene. However for our present objective we need an *expression vector*. This means that, upstream of our enzyme gene, the plasmid also carries special *promoter* sequences in the DNA that allow us, at a moment of our choosing, to switch on the translation of the gene to make large amounts of (i.e. express) the enzyme protein. Usually the 'switch' is a particular chemical substance. One first grows a large culture of healthy host cells (these may be *Escherichia coli* cells, but industrially yeast or other fungal hosts are often preferred) over several hours in a flask of nutrient solution, and only then adds the chemical signal. This diverts protein production to such an extent that, when the cells are broken open to 'harvest' the enzyme, it may make up as much as 50 per cent of all the protein (Figure 49).

In addition, molecular genetics companies have developed a range of clever expression vectors with additional DNA sequence to encode an extra protein tag on the end of the enzyme, giving it an extremely high affinity to a special chromatography column. One such tag is the bacterial protein *streptavidin* which attaches tightly to biotin. The chromatography therefore uses a column

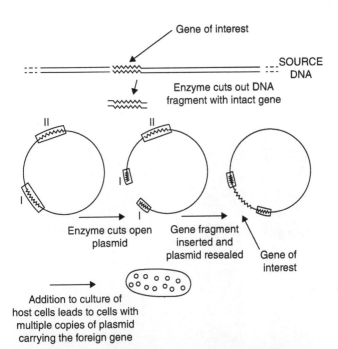

Gene of interest

Enzyme cuts out DNA fragment with intact gene

SOURCE DNA

Enzyme cuts open plasmid

Gene fragment inserted and plasmid resealed

Gene of interest

Addition to culture of host cells leads to cells with multiple copies of plasmid carrying the foreign gene

48. **Gene cloning.** The gene of interest is cut from source DNA using one or more restriction enzymes (Chapter 9). The same enzyme(s) cut(s) open the circular plasmid vector molecules allowing the gene to be inserted, re-forming the circle. The resealed plasmids enter the host cells and replicate. Typically the plasmid carries two or more genes for resistance to different drugs, and the cloning sites are in these genes (I and II in the diagram). Cells carrying the cloned gene will resist drug II but will be killed by drug I as the resistance gene has been cut. This allows selection of the correct cells with the cloned gene.

bristling with immobilized biotin, like baited hooks for the protein 'fish'. This type of strategy makes it possible often to achieve close to 100 per cent yield of a pure protein in a single step. The tag can then usually be removed with a proteinase.

Turning to the fragility issue, this problem arose in part because human investigators had concentrated on the enzymes of humans

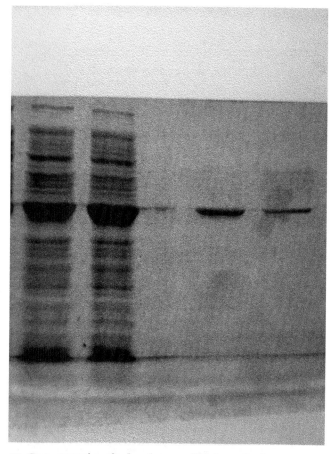

49. **Over-expression of a cloned enzyme. This is a stained electrophoresis gel similar to the one in Figure 12 (Chapter 3). The two left-hand tracks show massive over-expression of a bacterial enzyme cloned in *E. coli*. The right-hand tracks show the purified enzyme.**

and/or our close relatives, together with organisms like *E. coli* that normally grow inside mammals. Thus, for a long time, most attention was given to enzymes that had evolved to function at or around 37°C and did not need to be stable at higher temperatures.

However, from the 1960s onwards there was growing interest in *extremophiles*, organisms that live in extreme environments—salty, acidic, alkaline, cold, or hot. In recent times great effort has gone into studying the organisms that grow in 'black smokers'. Around these submarine hydrothermal vents the seawater is boiling under high pressure and at temperatures of 120°C or higher, yet there are organisms that thrive in this scalding broth. This means that all their molecules, nucleic acids, proteins, etc., must be able to survive these temperatures. This is partly thanks to special stabilizing substances that these organisms produce, but above all it is due to structural adaptations of the macromolecules themselves. Not surprisingly then, enzymes purified (or cloned) from thermophilic organisms are extraordinarily robust and tolerant to heat, and they offer great promise for industrial application.

Enzymologists today therefore expect to be able to obtain sturdy enzymes in amounts that make them economically realistic for industrial use, and supplying these enzymes is the central business of a number of large companies.

Enzymes for washing

One of the largest markets for enzymes is as additives in washing powders for clothes, tablets for dishwashers, etc. Traditionally, clothes and dishes would be washed with soaps and detergents that simply strip the grease and lift dirt off surfaces. Some kinds of soiling, however, are resistant under these conditions and enzymes offer potent assistance even though packet labels announcing 'New Bioactive' or 'With Added Enzymes' never explain what the enzymes actually do. Two types of enzyme are added to shift food stains. One is a *subtilisin* (see Chapter 6), a proteinase, and the other is a *lipase* which breaks down fats to more water-soluble products. These enzymes clearly need to resist the action of detergents, normally seen as among the most effective agents for unfolding protein molecules. The enzymes may also need to survive bleaching agents. In the past, the enzymes would also need to work at and resist high temperature, although, with the aid of enzymes, washing at 30–40°C is now recognized to be effective. As well as the enzymes, of course, the clothes need to survive the wash, and therefore, with a proteinase in the powder, one needs to remember that silk and wool are both proteins.

A third enzyme often added as a brightening agent for coloured cotton fabrics needs a little more explanation. *Cellulase* catalyses breakdown of cellulose, the substance that makes up cotton fibres. The theory here is that, with wear and washing, some of the tightly packed spun fibres come loose, forming a surface fuzz. This scatters light, so that intense colours look 'washed out'. If an enzyme cuts off the loose fibres, giving a fresh, fuzz-free surface, the garment should look brighter again. The snag the makers neglect to point out is that the more you do this the thinner your clothes become. Needless to say, this enzyme has no effect whatsoever on artificial, non-biological fabrics which are also entirely resistant to the other enzyme additives.

Enzymes to make food

The production and ripening of cheeses depends heavily on enzyme action. Traditionally the process begins with the separation of curds from whey promoted by the action of *rennin*, also known as chymosin, from calf stomach. Recent demand for vegetarian rennin has led to the cloning of the gene for the animal enzyme to be expressed in a microbial host. For the food industry this host organism cannot just be a microbiological favourite such as *E. coli*. It has to be one of a small number of approved *GRAS organisms* ('Generally Regarded as Safe' for the food industry).

The separated curds are used for the cheese ripening process, employing different living bacterial and/or fungal cultures for different varieties of cheese. Ripening takes many months and much of the flavour development is due to breakdown of *casein*, the major milk protein, into small, flavoursome peptides by the bacterial/fungal proteinases and *peptidases*. For some cheaper mass-produced cheeses the biological ripening process is nowadays speeded up by adding industrially produced peptidases.

Pectinases are another class of enzyme important in the food industry. Pectin, a carbohydrate arising from plant cell walls, is responsible, for example, for the cloudiness of crude fruit juices. Pectinase, by breaking pectin down to smaller sugars, clears the juice, improving consumer acceptance. Another enzyme with a special function in preparation of fruit juice is *glucose oxidase*. This catalyses the reaction of glucose, already present in the juice, with oxygen, and in this case the enzyme is added in order to mop up any dissolved oxygen to improve the long-term keeping properties of the juice once sealed.

An enzyme application that became economically important, though perhaps ethically dubious, was the production of *corn starch syrup*. In the 1970s there was a growing 'grain mountain' in North America and the European Union. Despite huge demand from starving populations elsewhere in the world, it was deemed uneconomical to move the mountain to the empty mouths. Instead, much effort went into converting the grain into a relatively high-value product closer to home. *Starch*, the storage carbohydrate in grain crops, is made of multiple linked glucose units. The enzyme *amylase* breaks it down, releasing glucose and forming a syrup. The plan was to sell corn syrup to the food industry to make biscuits, bread, cakes, ice cream, soft drinks, etc. However, a second enzyme greatly increased the profitability of this venture. Glucose is sweet, but the related 6-carbon sugar, *fructose* (Figure 6), is about three times sweeter, and the enzyme *glucose isomerase* interconverts these two sugars. This results in an equilibrium mixture rather than complete conversion, but the mixture is much sweeter than pure glucose. Raising the temperature pushes the equilibrium further towards fructose, and also a syrup is much more manageable for pouring or stirring

when it is warm. Heat stability is therefore essential for the
enzyme in this process.

The ethical dimension relates to the effort the food industry
put, on both sides of the Atlantic, into major marketing of
over-sweetened foods and also into discrediting scientific evidence
pointing to fructose possibly being harmful as a major dietary
component. In the accompanying campaign to vilify fat as the
truly harmful element of our diet, a bag of sugar could proudly
announce itself as '0% fat'. The result, thirty years on, is an
epidemic of obesity and diabetes, assisted, of course, also by
increasingly inactive lifestyles. The enzymes did their job but,
arguably in this case, a job better not done.

Skin, hair, and feathers

Another area of application for proteinases is in hair removal.
Apart from a coating of natural grease, hair, perhaps surprisingly,
is protein and, specifically, a protein called *keratin*. As we saw in
Chapter 5, proteinases vary widely in their specificity, and, to
break down this tough, resistant protein, a specialized *keratinase*
is needed. This is produced by certain species of bacteria which
are encouraged to maximize keratinase production by growing
them on chopped feathers, which, like hair, are made of keratin.
This enzyme has two contrasting applications. The first is in
cosmetic depilatory creams. On a rather larger scale, keratinase is
also used in the dehairing of hides for the leather industry, a
process that formerly required aggressive chemical treatment in
notoriously smelly tanneries.

Enzymes for farming and waste treatment

Cellulase, mentioned above in the context of washing powders,
has an important application in a quite different sphere, namely
farming. Farmers ferment harvested hay to produce *silage* to feed
ruminant animals through the winter. Fermentation can be

speeded up by adding enzymes to break down the tough cell walls of grass, which are made of a network of cellulose fibres.

Awareness in recent years of environmental degradation through dumping of various forms of waste on a vast scale into watercourses, landfill, etc., has led to a realization that some of these waste products are potentially valuable resources, and that their exploitation and recycling can be facilitated by use of appropriate enzymes. Important examples are:

ligninase, which helps to break down timber waste and other resistant agricultural waste materials and is also used in the paper industry for processing wood pulp;

lipase, used to process waste fats from restaurants and other food outlets in the production of 'biodiesel'.

A more recent concern is the accumulation of plastic in the environment. The very properties that made it a wonder substance have turned it into a scourge, seemingly impossible to get rid of. However, there is a glimmer of hope in the discovery of bacteria that are able to grow on one of the major forms of plastic and therefore have enzymes able to attack this resistant substance. It remains to be seen whether large-scale production of the cloned enzymes or use of the intact bacteria proves to be the more viable approach.

Enzymes for chemistry?

Nearly all the enzyme applications listed above have something in common. They break down a substrate into something smaller and usually more soluble. These are *hydrolytic* reactions, that is, they use the water in which everything is dissolved or suspended as a second substrate. Biochemists have long dreamt of using enzymes more ambitiously as chemical tools, working in the opposite direction, to build things up, that is, for synthesis. Until recently these ambitions were not encouraged by the chemists. For a traditional chemist, reaching for an enzyme seemed like cheating, an admission of intellectual defeat. Also, there were very real practical objections, starting with the drawbacks of cost and fragility mentioned earlier. Next, the exacting specificity of enzymes, which seems one of their remarkable virtues to a biochemist, is a considerable disadvantage for a chemist, who would prefer a catalyst to be versatile and useable with a wide range of alternative substrates. Chemists also work with many compounds that are insoluble in water and do so by resorting to organic solvents like acetone, methanol, hexane, acetonitrile, etc. It was assumed that, for many applications, this ruled out enzymes, which normally operate in a watery environment. However, a Russian scientist, Alexander Klibanov, working at the Massachusetts Institute of Technology, attempted the 'impossible', experimenting with enzymes in the most apparently hostile solvents. It turns out that in solvents least like water in their properties, such as hexane, these very properties result in a thin shell of water adhering to the protein molecules. This preserves them in their properly folded state so that they continue to work properly. A second important development has been the realization, turning again to extremophiles, that enzymes from very salt-tolerant organisms (*halophiles*) are also very tolerant of organic solvents. These offer promise for the future. Issues of cost, fragility, and solvent tolerance are no longer an insurmountable barrier to the use of enzymes for chemistry and there are already many successful examples of their routine application.

If we really want to extend the use of enzymes for industrial chemistry, however, we must also confront the issue of restricted or inappropriate specificity. There are enzymes in Nature to catalyse most of the types of reaction that chemists use, but more often than not they deal with substrates that would not interest the chemist. For instance, in organic synthesis an important task is joining two molecules to make a larger one by forming a new carbon-to-carbon chemical bond. The enzyme aldolase, met briefly in Chapter 5 (see Figure 34), catalyses just such a reaction (what chemists would call an 'aldol condensation'). However, an enzyme that only works with one or two sugar phosphates is useless for industrial chemistry if the chemist's target molecules are very different. This might once have seemed an insurmountable obstacle, but we shall see in the next chapter how molecular genetics is offering ways to solve even this problem.

Chapter 9
Enzymes and genes—new horizons

Tailored enzymes—a possibility?

As we saw earlier, the biological route to making new DNA relies on a single strand serving as template for a new growing strand, following the rules of G–C and A–T pairing. This can be done in the test tube with purified enzymes. If one introduces a 'primer', a short piece of synthetic DNA (around twenty units long) made to perfectly partner a stretch of the gene sequence, this will sit on the template DNA, forming a short stretch of double-stranded DNA. If the four necessary DNA building blocks and the enzyme *DNA polymerase* are added, the short double-stranded stretch will be extended by adding DNA bases one by one until the entire template has been converted to double-stranded DNA. Early on in the development of this methodology, however, it was realized that one could fool the system. Provided the primer was long enough, with plenty of perfect base pairing on either side, one could get away with a mismatch in the middle, say putting in an A instead of the expected C opposite a G in the template sequence (Figure 50). This *mutagenic primer* would then be extended to make a mutated copy of the gene. By designing the right synthetic primer (made on a programmable automated machine), one can introduce any desired mutation (out of the nineteen possibilities allowed by the genetic code) at any desired position in a cloned gene. It is also possible either to insert or to delete amino acids by

```
         Pro   Tyr   Lys   Gly   Gly   Leu
5'---GA│CCG│TAC│AAG│GGC│GGC│TTA---3'
      │ │ │ │ │ │ │ │ │ │ │ │ │ │ │ │
3'---CT│GGC│ATG│TTC│CCG│CCG│AAT---5'

   5'GA│CCG│TAC│CTG│GGC│GGC│TTA3'
      │ │ │ │ │ │ │═│ │ │ │ │ │ │ │
3'---CT│GGC│ATG│TTC│CCG│CCG│AAT---5'
```

50. Mismatch priming. The upper paired sequences show correct base pairing in a segment of the gene for a bacterial enzyme, glutamate dehydrogenase. Codons are indicated with the amino acids they encode. The codon for an active-site lysine is highlighted. In the bottom pair, the upper strand is a twenty-base synthetic mismatch primer successfully used to change the enzyme's substrate specificity. Changing <u>AAG</u> to <u>CT</u>G changes positively charged lysine to uncharged leucine.

having several adjacent mismatches on one strand or the other. This method of '*site-directed mutagenesis*' opened the door to what rapidly became known as *protein engineering*.

Initially people wondered whether mutated proteins would fold up properly. Occasionally they did not; but, overall, protein engineering has been remarkably successful and has become a powerful tool for pure research, allowing us to test the contribution of different parts of a protein structure with surgical precision. But let us focus here on a practical question: is it possible purposefully to alter an enzyme's substrate specificity?

Site-directed mutagenesis

The first successful engineered changes in enzyme active sites were announced at the start of the 1980s, by two research groups, one in a British academic laboratory, the other in an American biotech company. The UK project was a collaboration between three Cambridge scientists with complementary skills. They chose an unlikely enzyme in terms of immediate 'usefulness', but the objective was simply to ask 'Can it be done?' and to ask this for an enzyme that offered the prospect of an answer. The guiding philosophy was 'knowledge-based'. If one understood the enzyme well enough, and had a detailed 3-D structure and the cloned gene, then one had the essential minimum requirements for the task. The chosen enzyme, *tyrosyl tRNA synthetase*, is one of the enzymes that read the genetic code and prepare the twenty amino acids to be built into proteins (Chapter 4). The enzymologist, Alan Fersht, who had studied this enzyme in depth, collaborated with the X-ray crystallographer David Blow, who solved the structure, and the molecular geneticist Greg Winter who contributed the DNA skills. Having first shown that targeted mutations could be made and the proteins successfully expressed and purified, the team went on to make systematic mutations at various positions in the active-site region of this enzyme, drawing out lessons, not just about how tyrosyl tRNA synthetase works and what governs its very tight substrate specificity, but also more generally about the individual contributions of particular physical interactions.

Independently, in California, a team at Genentech, one of the new biotech companies, picked an enzyme that would pose similar questions but might also deliver commercially useful outcomes if the new approach really worked. Their chosen enzyme was *subtilisin*, the washing powder proteinase, and they addressed genuine practical concerns in making enzyme-based powders, such as how to resist the damaging effect of bleach on proteinase activity. Again the approach was knowledge-based, but with one revealing difference. At this stage each new mutant required a separate mutagenic primer. These were made to order by a commercial supplier, but each one then cost about £500. Academic groups on tight budgets had to think carefully about how many mutations to make at a chosen position. Typically the Cambridge group restricted themselves to two or three likely choices. This restriction, of course, then depends heavily on how well one truly understands the structure. The Genentech study offers a striking contrast. Subtilisin loses activity in the presence of bleach owing to chemical modification of one particular amino acid, No. 222 in the overall sequence, a methionine residue. The hope, therefore, was that, assuming this methionine was not absolutely essential for the activity, it could be replaced by another amino acid that was not damaged by bleach. But which? A prudent academic group would certainly have selected one or two amino acids similar to methionine in size and shape. The Genentech team faced the truth that they could not confidently guess which substitution would be best, and decided that, for a legitimate commercial target, they should spend what they needed to spend, and so they bought nineteen different primers in order to make every possible amino acid change. The interesting and unexpected result was that the best mutations were not the ones that introduced another amino acid of similar size and shape. This offered a cautionary lesson for all enzymologists not to over-estimate the depth of their understanding.

In the British context, it is interesting to note that, at this early stage, industry was prepared to take a direct interest in supporting purely academic research like the Cambridge project and many others that soon followed elsewhere. A joint public–private partnership supported a number of such projects. In a sense, it was a positive advantage for the projects not to be close to commercial interests, so that rival companies could sit together round the table, all interested in finding out whether the method worked. Once it became clear that it did, they could disappear behind closed doors to apply protein engineering to their own commercial targets.

Random mutagenesis and screening

Since the early 1980s there have been rapid advances in the methodology of molecular genetics. Robotic techniques now allow experiments on a much larger scale, and, crucially, prices of essential ingredients, such as synthetic primers, have plummeted. This has made possible a philosophy of protein engineering almost diametrically opposite to the one discussed above. There are two variables to consider: first, which amino acid(s) to change; and, second, what change(s) to make. As site-directed mutagenesis has been shown to 'work' again and again for hundreds of different enzymes, scientists have become more and more adventurous in selecting targets, seeking to change heat stability, pH dependence, etc., and having to recognize that often they had very little idea what change to make. In this situation it makes a lot of sense, instead of committing to a small number of carefully chosen mutations, to adopt a scattergun approach, making random mutations at many positions. This inevitably generates a very large number of cells to screen. Nineteen possible mutations at, say, 200 positions means nearly 4,000 possibilities, and, to be sure of catching at least one copy of each, it is probably necessary to screen ten times that number. Each of the mutant cells has to be grown up to produce enough cells for a functional test to see whether that particular mutation is beneficial. The vast majority will be discarded, but, with luck, a small number can be selected for more detailed investigation (Figure 51).

This description implies going from one extreme (totally knowledge-based and targeted) to the other (totally random). Even with robots and cheaper materials, a totally random approach generates much work and expense, and so, in practice, the best strategy may be a middle way. In terms of the amino acids to mutate, it is often possible to narrow down the likely influential positions from a few hundred to a dozen or so. Also it is possible to restrict the number of mutations allowed at each position to something less than the full nineteen (on the basis that some amino acids are quite similar and can perhaps deputize for one another). There is nevertheless always a risk of missing something good by cutting too many corners.

Real targets: left- and right-handed molecules

Today's protein engineers thus have a choice of techniques and have to select according to a sober appraisal of how well they

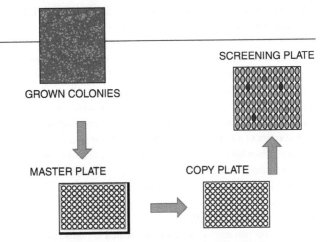

51. **Plate screening. After random mutagenesis cells will all carry a copy of the cloned gene. Some may be unmutated. Others will carry mutations that offer no advantage. The objective is to find any offering an improvement. On the first plate cells are spread out so that each produces a separate colony of many cells. Using a robot, individual colonies are picked into numbered wells on a master plate. This is used to seed copy plates. Cells grown on a copy plate are tested for the desired new activity on a screening plate so that beneficial mutations can be identified and saved.**

really understand their enzyme. As always in science, there are many mistakes and many failures but also many bull's eye hits and also lucky breaks. Over nearly forty years, we have learnt a great deal about which objectives are easy to achieve and which are more difficult. Fortunately, substrate specificity is one of the easier objectives to target. This is because, with a high-resolution 3-D structure for the enzyme, ideally with the substrate or a substrate analogue bound at the active site, one can decide with some confidence which small number of amino acid sidechains determine the enzyme's preference for its substrate. There are thus now numerous examples of successful remodelling of active sites to change or extend substrate specificity. Rather than making a lengthy list, I will describe one area of application in which I have been directly involved.

In order to do so, I must first introduce the concept of handedness (*chirality*) in chemical molecules. It arises from the fact that atoms are 3-D. With four valency arms (Chapter 2), a carbon atom can be attached to four different partners. The four bonds are symmetrically arranged in space, so that they point towards the four corners of a tetrahedron (Figure 52). If the four partners are different, then there are two ways they can be arranged, giving two different mirror-image structures. A biochemically relevant example is that of amino acids. Apart from glycine, which has two hydrogen atoms occupying two of its four positions, every other α-amino acid has four different substituents. Thus alanine, for example, has two mirror-image forms which are referred to as L-alanine and D-alanine. As it happens, Nature has chosen to use only L-amino acids (plus glycine) in building proteins.

Molecular handedness is important for pharmaceuticals. The industry is permanently scarred by a scandal from the 1950s and 1960s. A drug called *thalidomide* was marketed to control morning sickness in pregnancy. Though very successful in helping expectant mothers, it unfortunately also affected the development

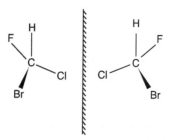

52. **Chiral carbon atom. The four valencies of a carbon atom are symmetrically directed in 3-D space towards the four corners of a tetrahedron. If the C is bonded with four different substituents as above, there are two mirror-image versions that are not superposable.**

of the foetus, resulting in babies born with arms or legs partly or wholly missing. This drug was a 50:50 mixture of two mirror-image molecules (a so-called *racemic* mixture)—because chemical synthetic reactions tend to be even-handed. Ultimately it emerged that morning sickness was cured by one version of the molecule and the birth defects were caused by the other. Ever since the thalidomide disaster there has been strong pressure in the development of new drugs either to use chirally pure components (i.e. avoid racemic mixtures) or to prove beyond reasonable doubt that the 'wrong' mirror-image partner is harmless.

This highlights a huge advantage of enzyme catalysis. In contrast to the even-handed traditional chemical catalysis, an enzyme dealing with its natural substrate will display absolute discrimination between the mirror-image forms. This is not only easy, it is more or less inevitable, since the enzyme molecule itself is an asymmetric structure made up of asymmetric amino acids.

Chemistry does have methods for separating ('resolving') the two partners in a racemic mixture, but such labour-intensive tasks are better avoided. One response from the chemists has been to copy biology. This biomimetic catalysis relies on constructing a synthetic catalyst that, just like an enzyme, presents an asymmetric surface that will discriminate between racemic partners. These approaches are ingenious but the level of discrimination never comes near what Nature has achieved through millions of years of natural selection. The biochemist's inevitable thought is, 'Since enzymes do the job better, why not use them?'

Amino acid dehydrogenases

The amino acid dehydrogenases are a family of enzymes that either remove the amino group ($-NH_2$) from an L-amino acid, initiating its breakdown, or, in the opposite direction, introduce the amino group to make the amino acid. The best-studied and most widely found member of the family is glutamate

L-glutamate   2-oxoglutarate

$$COO^-$$
$$|$$
$$CH_2$$
$$|$$
$$CH_2 \quad + \boxed{NAD^+} \rightleftharpoons$$
$$|$$
$$HC\ \overset{+}{N}H_3$$
$$|$$
$$COO^-$$

$$COO^-$$
$$|$$
$$CH_2$$
$$|$$
$$CH_2 \quad + \boxed{NADH} + H^+$$
$$|$$
$$C = O \qquad + NH_4^+$$
$$| \qquad\qquad \text{ammonium ion}$$
$$COO^-$$

53. Reversible oxidative deamination catalysed by glutamate dehydrogenase. NAD$^+$ is biochemists' shorthand for the oxidative coenzyme nicotinamide adenine dinucleotide. This widespread biological reaction moves nitrogen (as ammonia) in and out of organic combination.

dehydrogenase (GDH), which catalyses the reaction in Figure 53, interconverting L-glutamic acid and 2-oxoglutarate.

Essential for our story is the fact that 2-oxoglutarate is not a chiral compound. In particular, the α-carbon atom (see Chapter 3) does not have four different attachments. On the other hand, the catalytic reaction introducing the amino group creates a handed molecule, an L-amino acid. Around 1990, our team at the University of Sheffield obtained the first 3-D structure for a GDH and cloned the gene (molecular genetic expertise contributed by John Guest, crystallography by David Rice, and enzymology by myself). The crystallography told us exactly where and how the substrate L-glutamic acid sits in the active site of the enzyme. At one end of the substrate, the enzyme provides a positive charge (L-lysine) to anchor the negatively charged α-carboxyl group, common to every α-amino acid. At the other end it provides another lysine positive charge (K89) and a second sidechain (L-serine, S380) with an –OH group, offering a snug welcome for the sidechain carboxyl group of the substrate (Figure 54).

In black: Substrate glutamate
In dark grey: Key enzyme sidechains
NIC indicates nicotinamide portion of substrate NAD+
This is a schematic 2-D sketch of a complex 3-D object. Real
3-D coordinates can be found in the international protein database.

54. **Glutamate dehydrogenase active site: amino acid sidechains that interact with the substrate glutamate. When glutamate arrives, K89 and S380 provide recognition and binding for the sidechain carboxylate and K113 anchors the α-carboxylate. The enzyme hinges, enclosing the active site and establishing the geometry for oxidation by NAD+ catalysed by D165 and K125.**

It is this second recognition site that is critical for our mission, since this is what distinguishes GDH from any other amino acid dehydrogenase. The challenge, therefore, was to use this knowledge to change the amino acid substrate specificity. Sure enough, mutating the two positions just mentioned to amino acids with hydrophobic sidechains, plus one other position, delivered a new

55. Methionine and glutamic acid. The methionine sidechain is slightly longer and lacks the negative charge.

enzyme that was no longer active with L-glutamate but was now active with the hydrophobic amino acid L-methionine (Figure 55).

Meanwhile, Japanese scientists were publishing gene sequences for other members of the enzyme family, dehydrogenases for L-phenylalanine (PheDH), for L-leucine, and for L-valine. We realized that these were indeed family members, showing amino acid sequence similarities (see Chapter 6) (Figure 56). These similarities were sufficient to see that all the features of GDH that one would expect to see in any L-amino acid dehydrogenase were present in every one of the sequences.

Crucially, however, the two sidechains responsible for glutamate specificity were replaced by hydrophobic amino acids in all these other dehydrogenases. This gave the crystallographers the confidence to build the sequence of PheDH into the solved high-resolution sequence of GDH. In other words, this was a guessed structure for PheDH based on the assumption that the overall structures would be closely similar. Guided by this, we took the risky step of mutating residues in the assumed active site of PheDH. This worked beyond expectation and allowed us to

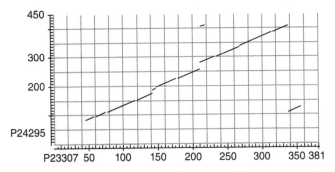

56. DIAGON sequence comparison of glutamate dehydrogenase and phenylalanine dehydrogenase. The program provides an optimal alignment of similar sequences, inserting occasional gaps if necessary. These show up as breaks in the continuous diagonal.

manipulate the specificity of PheDH. By introducing mutations at two or three critical positions in the predicted active site we were able to widen the range of amino acids the enzyme would accept as substrates. In particular, a number of non-biological amino acids worked several hundred times faster with the mutant enzymes than with the original PheDH. These non-biological amino acids are in great demand by the pharmaceutical industry and command a high price since the chirally pure L- or D-forms are difficult to make by chemical synthesis. Our engineered enzymes could now be used in two ways (Figure 57) to make these high-value compounds.

If the non-chiral precursor equivalent to 2-oxoglutarate in Figure 53 could be made chemically, then the enzyme would simply catalyse the reaction to make the chirally pure L-amino acid. If, on the other hand, it was easier chemically to make the amino acid, this would be available as the racemic mixture of equal amounts of L- and D-. In this case, the enzyme could be used to catalyse the reaction in the opposite direction. This would remove all the L-amino acid, converting it to the non-chiral precursor, leaving the D-amino acid untouched. Simple separation methods would give the pure D-amino acid and then the enzyme

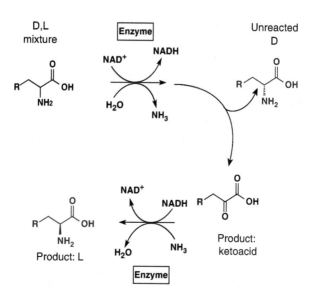

57. Chiral synthesis/resolution with an amino acid dehydrogenase.

could be used in a second reaction to convert the separated precursor to the L-amino acid.

More recently, protein engineers at Georgia Tech in the USA have carried out an ambitious modification of the substrate specificity of another family member, leucine dehydrogenase, in a different direction. Could one perhaps alter the active site in such a way that the enzyme no longer required the substrate to have a carboxyl group at the alpha position? Such a compound would not be an amino acid but an *amine*. As it happens, industrial chemists are even more interested in chirally pure amines than in amino acids, and the US group found that simple mutations at the recognition site for the α-carboxyl group, removing the positive charge, delivered the desired new specificity.

Protein engineering has moved from a wary adventure on the part of a few pioneering groups in the 1980s to a routine tool used around the world today on a vast range of targets, and with immense power for fine-tuning the properties of enzymes and other proteins for practical applications. In particular, enzyme catalysis has at last gained widespread acceptance as a tool for chemistry, and, because enzyme catalysis avoids the need for high temperatures and pressures and toxic waste streams, it has become the basis of what is now generally known as 'green chemistry'. One recent indicator of this remarkable shift in the relationship between enzymology and mainstream chemistry was the share in the 2018 Nobel Prize for Chemistry for Professor Frances Arnold at Caltech who has in recent years pioneered major advances in the methodology for efficient development of novel biocatalysts.

Enzymes, biological defence, and the genetic revolution

Much of this chapter has taken for granted the techniques of modern molecular genetics. We have assumed that we can clone genes at will, determine their DNA sequences, express them to high levels in foreign host organisms, and go on to make changes that alter the functional properties of the corresponding protein. But how is this done? The answer, of course, is that it depends on a large number of very sophisticated enzymes. The details of DNA methodology would more than fill another book, but we should at least list here three breakthrough discoveries. Between them they have transformed biotechnology, biology, and our lives, but it is very important to realize that the discoveries were made purely by pursuing scientific curiosity, not because of the anticipated impact. Scientists all understand that if we knew what was around the next corner we would be there already. Unfortunately many of the people who make the funding decisions think that the goals are obvious, when in fact so many of the momentous advances are made almost by accident.

The first of these momentous advances, made by following up a seemingly esoteric problem, won Nobel Prizes in 1978 for the Swiss microbiologist Werner Arber and the Americans Hamilton Smith and Daniel Nathans There is a class of viruses called *bacteriophage* that make their living by infecting bacteria and often killing them. Back in the 1950s and 1960s microbiologists were puzzling over the apparently random pattern of infection, with certain species of bacteria able to resist infection by some bacteriophage but not others. It emerged that bacteria have a defence mechanism which relies on enzymes that cut up the invader's DNA. Why, then, were these bacteria not resistant to all bacteriophage? The answer lies in the remarkable specificity of these *restriction enzymes*, so called because they restrict the range of hosts available to the bacteriophage. We have seen that the enzymes that cut up proteins do so at positions determined by the amino acid sequence, and some of them require a very definite sequence of two, three, or four amino acids. Restriction enzymes are even fussier, often requiring a precise sequence of six DNA bases or in a few cases even eight, in order to cut. Different bacteria have different restriction enzymes with different DNA sequence requirements. *Escherichia coli*, for example, produces the restriction enzyme EcoR1 with a recognition sequence GAATTC (Figure 59). Viruses have only a small number of genes of their own and so it is quite possible for the sequence GAATTC not to occur even once in the genome of a particular bacteriophage. Which bacteriophage can successfully invade which bacteria will thus depend on what restriction enzymes the bacteria possess and on the actual DNA sequence of the bacteriophage. Since bacteria have much larger genomes than the invading viruses it might seem that they would be more vulnerable to their own restriction enzymes, but in fact they have an enzymatic system for protecting their own DNA at the vulnerable cut sites.

Once the remarkable specificity of restriction enzymes was recognized, they became powerful tools for molecular geneticists,

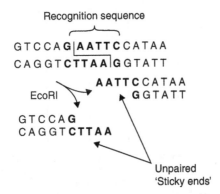

Recognition sequence

```
GTCCAG|AATTC|CATAA
CAGGTCTTAA|GGTATT
```

EcoRI

```
        AATTCCATAA
        GGTATT
```

```
GTCCAG
CAGGTCTTAA
```

Unpaired
'Sticky ends'

58. Restriction enzyme cleavage of double-stranded DNA by EcoR1.
The staggered pattern of cleavage generates 'sticky ends'.

enabling them to map DNA in terms of the distribution of cut sites for different restriction enzymes and allowing them, for cloning purposes, to cut down large pieces of DNA to fragments containing a gene of interest with very little extra DNA. One particular obliging feature of restriction enzymes is that, since many of them cut double-stranded DNA in a staggered fashion (Figure 58), they generate 'sticky ends' making it relatively easy to insert the excised gene into a plasmid vector cut with the same restriction enzyme.

The second transformative piece of enzymology, the inspired technique known as PCR (see Chapter 7), also won a Nobel Prize, in Chemistry, in 1993, for its inventor, Kary Mullis. This is a clever method for enormously amplifying a targeted DNA sequence, producing millions of identical molecules starting with very few molecules of the starting template. This method was developed and refined through the mid-1980s, initially at Cetus Corporation, an American biotech company where Mullis worked. The enzyme in question is DNA polymerase. As we have already seen, with a short primer to get it started, this enzyme will build double-stranded DNA on a single-stranded template.

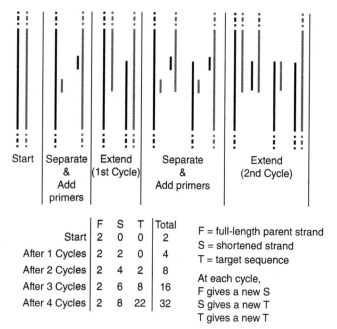

	F	S	T	Total
Start	2	0	0	2
After 1 Cycles	2	2	0	4
After 2 Cycles	2	4	2	8
After 3 Cycles	2	6	8	16
After 4 Cycles	2	8	22	32

F = full-length parent strand
S = shortened strand
T = target sequence

At each cycle,
F gives a new S
S gives a new T
T gives a new T

59. **Polymerase chain reaction. After two cycles the amplification of the target sequence rapidly builds up.**

If we consider a primer that attaches in the middle of a linear single-strand template, DNA polymerase activity would result in a molecule single-stranded up to the primer and then double-stranded for the rest of the length. To understand PCR we need to know that heating double-stranded DNA will separate the strands. Mullis's brilliant innovation was to introduce a second primer attaching to the other strand and priming extension in the opposite direction (Figure 59). If the two primers are either side of the segment containing the sequence of interest (e.g. a gene), then we can start with double-stranded DNA, heat it to generate two single-stranded templates, cool it to allow the primers to attach, and extend it, resulting in two shortened strands as well as the original full-length strands. Heating again results in full-length templates plus shortened templates. Once we go through another cycle of extension, we start to make pieces of DNA that extend only between the two primers, and since these generate short templates on heating, the result is that with every cycle one generates more and more of the short targeted section.

When the PCR method was initially developed, the drawback was that within each cycle the heating destroyed the DNA polymerase, so that for twenty cycles there would need to be twenty additions of fresh enzyme. Here extremophiles came to the rescue. The DNA polymerase of the bacterium *Thermus aquaticus*, isolated from a hot spring in Yellowstone Park, was stable up to the temperatures needed to separate the DNA strands, so that with this enzyme only one addition was necessary. Programmable heaters were rapidly developed able to look after multiple samples through many cycles without constant operator intervention.

It is difficult to overstate the impact of this technique. It gave a method for detecting HIV DNA, and more recently the DNA of the virus responsible for the COVID-19 pandemic, in patient blood samples. It was able to amplify DNA from a single starting molecule of DNA. A single sperm could be analysed. Likewise early-stage embryos. Ancient partially degraded DNA from early hominids could be amplified and meaningfully analysed, revolutionizing our understanding of our own evolution. Perhaps with most impact on public consciousness, it became possible to solve many crimes from tiny amounts of material left behind by a guilty party, even a single hair. And PCR primers can now be used, similarly to the methods mentioned earlier, to carry out mutagenesis with ease, by incorporating mismatches in the primer sequences.

Finally, in the new millennium we have another breakthrough with another remarkable enzyme, again associated with bacterial defence, in this case the 'CRISPR-Cas9' system. CRISPR stands for 'clustered regularly interspaced short palindromic repeats' and refers to an odd feature of the DNA in many bacterial cells first noted about thirty years ago. As the name implies, a region of the genome has a short sequence that is multiply repeated with other bits of sequence in between the repeats. It was some time before it was realized that these short sequences were captured from invading foreign DNA (plasmids or bacteriophage) and that they constitute a kind of molecular memory. This is similar to our own immune defence mechanisms, although those are protein-based rather than DNA-based. The CRISPR system is reawakened if there is repeat invasion by the same source of DNA. The CRISPR sequences become the template for producing RNA transcripts (see Chapter 5). The RNA attaches itself to Cas9, which is a DNA-degrading enzyme, and the part of the RNA transcribed from one of the CRISPR repeats guides Cas9 to the matching sequence in the invading DNA, which is then cut.

This clever system is, of course, a valuable defence for the bacteria, but recently has taken on a much wider significance. US and French scientists Jennifer Doudna and Emmanuelle Charpentier realized that Cas9 could be guided to other kinds of DNA, for instance that in crop plants or even in human embryos, by supplying suitably matching pieces of RNA. Once the targeted DNA is cut inside a living cell, there are natural enzymatic DNA repair mechanisms. These can be guided by a supplied DNA template and this means that the enzyme machinery can be used either to introduce a new mutation or to correct an undesirable mutation. This latter now opens up the possibility of correcting disease mutations such as those discussed in Chapter 6. Inevitably it will also raise profound new ethical and regulatory concerns.

As we have seen, enzymes are minute but powerful molecular machines, machines that carry out various sorts of chemistry at phenomenal speed and with extraordinary accuracy. Chapter 1 closed with a quotation from Frederick Gowland Hopkins, who suggested almost ninety years ago that it was 'difficult to exaggerate the importance to biology, and to chemistry no less, of extended studies of enzymes and their action'. A prophetic statement indeed, but even Hopkins could not have dreamt of the sweeping impact of these amazing molecules on science, technology, medicine, and our understanding of life itself.

References

Chapter 1: No enzymes, no life

William Henry (1826), quoted in J. W. Webster, On the basis of Professor Brande's chapter VI: analysis of vegetable substances, *A Manual of Chemistry*, Richardson & Lord, p. 456.

Friedrich Wöhler (1828), Ueber künstliche Bildung des Harnstoffs, *Annalen der Physik und Chemie*, 88, 253–6. English translation available at https://www.chemteam.info/Chem-History/Wohler-article.html

A. Payen and J. F. Persoz (1833), Memoire sur la diastase, les principaux produits de ses reactions et leurs applications aux arts industriels [Treatise on diastase, the principal products of its reactions and their application to industrial arts], *Ann. Chim. (Phys)*, 53, 73.

Chapter 2: Making things happen—catalysis

P. C. Engel (1981), *Enzyme Kinetics: The Steady-State Approach*, 2nd Edition, Chapman and Hall Outline Studies in Biology. A short introduction to a topic that students find off-putting because of the maths.

A. Cornish-Bowden (1995), *Fundamentals of Enzyme Kinetics*, Revised Edition, Portland Press. A well-written, deeper coverage of the same topic.

Chapter 3: The chemical nature of enzymes

J. B. S. Haldane ([1930] 1965), *Enzymes*, Reprinted by MIT Press. A fascinating account of how much was already known but how much was also not known about enzymes ninety years ago.

J. B. Sumner (1926), The purification and crystallisation of urease, *J. Biol. Chem.*, 69, 435.

J. H. Northrop, M. Kunitz, and R. M. Herriott (1948), *Crystalline Enzymes*, 2nd Edition, Columbia University Press.

Richard Willstätter ([1930] 1965), The purification and chemical nature of enzymes, in J. B. S. Haldane, *Enzymes*, Reissued by MIT Press. Gives an account of Willstätter's contribution and table XX lists the various enzymes he purified.

J. R. Knowles (1991), Enzyme catalysis: not different, just better, *Nature*, 350, 121.

F. Sanger and E. O. P. Thompson (1953), Determination of the amino acid sequence of bovine insulin, *Biochemical Journal*, 53, 366.

J. D. Watson and F. H. C. Crick (1953), Molecular structure of nucleic acids: a structure for deoxyribose nucleic acid, *Nature*, 171, 737–8. Surely one of the most understated breakthroughs in the history of science.

J. D. Watson (1968), *The Double Helix*, Simon and Schuster. A good deal less understated, Jim Watson's revealing personal account of the discovery.

Sir William Lawrence Bragg (1992), *The Development of X-Ray Analysis*, Dover Publications. Reprint of a 1975 volume by the son in the father and son collaboration that, in 1912, solved the first X-ray structure: that of crystalline sodium chloride (a structure dismissed at the time as 'chessboard evidence' by H. E. Armstrong).

J. C. Kendrew and R. G. Parrish (1957), The crystal structure of myoglobin III: Sperm whale myoglobin, *Proceedings of the Royal Society of London* A, 238, 305.

J. C. Kendrew, G. Bodo, H. M. Dintzis, R. G. Parrish, H. Wyckoff, and D. C. Phillips (1958), A three-dimensional model of the myoglobin molecule obtained by X-ray analysis, *Nature*, 181, 662.

L. Pauling, R. B. Corey, and H. R. Branson (1951), The structure of proteins: two hydrogen-bonded helical configurations of the polypeptide chain, *Proceedings of the National Academy of Sciences U.S.A.*, 37, 235.

R. H. Pain (ed.) (1994), *Mechanisms of Protein Folding*, Oxford University Press. Describes prevailing concepts and experimental approaches.

V. Daggett and A. R. Fersht (2003), Is there a unifying mechanism for protein folding?, *Trends in Biochemical Science*, 28, 18.

C. B. Anfinsen (1973), Principles that govern the folding of protein chains, *Science*, 181, 123. An article by one of the pioneers in the field of protein folding.

Chapter 4: Structure for catalysis

C. C. F. Blake, L. N. Johnson, G. A. Mair, A. C. T. North, D. C. Phillips, and V. R. Sarma (1967), Crystallographic studies of the activity of hen egg-white lysozyme, *Proceedings of the Royal Society of London* B, 167, 378.

D. M. Chipman and N. Sharon (1969), Mechanism of lysozyme action, *Science*, 165, 454.

L. N. Johnson and G. A. Petsko (1999), David Phillips and the origin of structural enzymology, *Trends in Biochemical Science*, 24, 287.

D. E. Koshland Jr (1958), Application of a theory of enzyme specificity to protein synthesis, *Proceedings of the National Academy of Sciences U.S.A.*, 44, 98.

R. Wolfenden (1969), Transition state analogues for enzyme catalysis, *Nature*, 223, 704.

K. Dalziel (1957), Initial steady state velocities in the evaluation of enzyme-coenzyme-substrate reaction mechanisms, *Acta Chemica Scandinavica*, 11, 1706. A paper that showed how much information about how enzymes assemble their substrates can be gained from systematic measurement of reaction rates.

S. J. Benkovic and S. Hammes-Schiffer (2003), A perspective on enzyme catalysis, *Science*, 301, 1196.

M. C. Linder (ed.) (1991), *Nutritional Biochemistry and Metabolism*, 2nd Edition, Elsevier.

Chapter 5: Enzymes in action

C. de Duve (1963), The lysosome, *Scientific American*, May. An article by the Belgian Nobel Laureate whose use of the centrifuge and enzymology revealed two previously unknown organelles in mammalian cells.

A. Hershko and A. Ciechanover (1998), The ubiquitin system, *Annual Review of Biochemistry*, 67, 425.

G. N. de Martino and T. G. Gillette (2007), Proteasomes: machines for all reasons, *Cell*, 129, 659.

A. Varshavsky (2005), Regulated protein degradation, *Trends in Biochemical Science*, 30, 286.

J. F. Kerr, A. H. Wyllie, and A. R. Currie (1972), Apoptosis: a basic biological phenomenon with wide-ranging implications in tissue kinetics, *British Journal of Cancer*, 26, 239.

D. R. McIlwain, T. Berger, and T. W. Mak (2013), Caspase functions in cell death and disease, *Cold Spring Harbor Perspectives in Biology*, 5, a008656.

R. M. Lawn and G. A. Vehar (1986), The molecular genetics of hemophilia, *Scientific American*, 254, 48.

M. M. Krem and E. di Cera (2002), Evolution of cascades from embryonic development to blood coagulation, *Trends in Biochemical Science*, 27, 67.

F. H. C. Crick (1966), Codon-anticodon pairing: the wobble hypothesis, *Journal of Molecular Biology*, 19, 548.

M. Ibba and D. Söll (2000), Aminoacyl-tRNA synthesis, *Annual Reviews of Biochemistry*, 69, 617.

J. Ling, N. Reynolds, and M. Ibba (2009), Aminoacyl-tRNA synthesis and translational quality control, *Annual Reviews of Microbiology*, 63, 61.

C. L. Markert, J. B. Shaklee, and G. S. Whitt (1975), Evolution of a gene: multiple genes for LDH isozymes provide a model of the evolution of gene structure, function and regulation, *Science*, 189, 102.

C. C. Rider and C. B. Taylor (1980), Isoenzymes, *Outline Studies in Biology*, Chapman & Hall.

Chapter 6: Metabolic pathways and enzyme evolution

W. Eventhoff and M. Rossmann (1976), The structures of dehydrogenases, *Trends in Biochemical Science*, 1, 227.

M. S. Johnson and J. P. Overington (1993), A structural basis for sequence comparisons: an evaluation of scoring methodologies, *Journal of Molecular Biology*, 233, 716.

E. Zuckerkandl and L. Pauling (1965), Molecules as documents of evolutionary history, *Journal of Theoretical Biology*, 8, 357.

Enzymes

H. Neurath (1984), Evolution of proteolytic enzymes, *Science*, 224, 350.

C. A. Orengo, D. T. Jones, and J. M. Thornton (2003), *Bioinformatics: Genes, Proteins and Computers*, BIOS Scientific Publishers.

N. H. Horowitz (1945), On the evolution of biochemical syntheses, *Proceedings of the National Academy of Sciences U.S.A.*, 31, 153.

T. R. Cech (1986), RNA as an enzyme, *Scientific American*, 255, 64.

C. Carola and F. Eckstein (1999), Nucleic acid enzymes, *Current Opinion in Chemical Biology*, 3, 274.

E. A. Doherty and J. A. Doudna (2000), Ribozyme structures and mechanisms, *Annual Review of Biochemistry*, 69, 597.

Chapter 7: Enzymes and disease

C. R. Scriver, A. L. Beaudet, W. S. Sly, D. Valle, B. Childs, K. W. Kinzler, and B. Vogelstein (eds.) (2001), *The Metabolic and Molecular Bases of Inherited Disease*, 8th Edition, McGraw-Hill. A vast multi-volume compendium of fascinating articles on the biochemistry and enzymology of individual hereditary enzyme defects. Now also available online.

H. Erlandsen and R. C. Stevens (1999), The structural basis of phenylketonuria *Molecular Genetics and Metabolism*, 68, 103.

A. E. Garrod (1963), *Inborn Errors in Metabolism*, Oxford University Press. This is a reprint of Garrod's classic 1909 work with a supplement by Professor H. Harris.

A. Wenz, C. Thorpe, and S. Ghisla (1981), Inactivation of general acyl-CoA dehydrogenase from pig kidney by methylenecyclopropyl acetyl-CoA, a metabolite of hypoglycin, *Journal of Biological Chemistry*, 256, 9809.

E. J. Corey, B. Czakó, and L. Kürti (2007), *Molecules and Medicine*, Wiley. An account of the development and mode of action of numerous drugs.

E. Ravina (2011), *The Evolution of Drug Discovery: From Traditional Medicines to Modern Drugs*, Wiley.

M. A. Stahmann, C. F. Huebner, and K. P. Link (1941), Studies on the hemorrhagic sweet clover disease. V: identification and synthesis of the hemorrhagic agent, *Journal of Biological Chemistry*, 138, 513.

M. Wainwright (1990), *Miracle Cure: The Story of Penicillin and the Golden Age of Antibiotics: Story of Antibiotics*, Wiley-Blackwell.

C. G. Smith and J. R. Vane (2003), The discovery of captopril, *FASEB Journal*, 17, 788.

M. Patlak (2004), From viper's venom to drug design: treating hypertension, *FASEB Journal*, 18, 421.

A. G. Tomaselli, S. Thaisrivongs, and R. L. Heinrikson (1996), Discovery and design of HIV protease inhibitors as drugs for treatment of AIDS, *Advances in Antiviral Drug Design*, 2, 173.

A. Wlodawer (2001), Rational approach to AIDS drug design through structural biology, *Annual Reviews of Medicine*, 53, 595.

Chapter 8: Enzymes as tools

P. Gacesa and J. Hubble (1987), *Enzyme Technology*, Open University Press.

J. Tramper and Y. Zhu (2011), *Modern Biotechnology: Panacea or New Pandora's Box*, Wageningen Academic Publishers.

J. Whittall and P. W. Sutton (eds.) (2010), *Practical Methods for Biocatalysis and Biotransformations*, Wiley.

R. N. Patel (ed.) (2000), *Stereoselective Biocatalysis*, CRC Press.

Chapter 9: Enzymes and genes—new horizons

G. Winter, A. R. Fersht, A. J. Wilkinson, M. Zoller, and M. Smith (1982), Redesigning enzyme structure by site-directed mutagenesis: tyrosyl tRNA synthetase and ATP binding, *Nature*, 299, 756.

A. R. Fersht (1987), Dissection of the structure and activity of the tyrosyl-tRNA synthetase by site-directed mutagenesis, *Biochemistry*, 26, 8031.

D. A. Estell, T. P. Graycar, and J. A. Wells (1985), Engineering an enzyme by site-directed mutagenesis to be resistant to chemical oxidation, *Journal of Biological Chemistry*, 260, 6518.

F. H. Arnold and G. Georgiou (eds.) (2010), *Directed Enzyme Evolution: Screening and Selection Methods*, Humana Press.

P. J. Baker, K. L. Britton, P. C. Engel, G. W. Farrants, K. S. Lilley, D. W. Rice, and T. J. Stillman (1992), Subunit assembly and active site location in the structure of glutamate dehydrogenase, *Proteins: Structure, Function and Genetics*, 12, 75.

J. K. Teller, R. J. Smith, M. J. McPherson, P. C. Engel, and J. R. Guest (1992), The glutamate dehydrogenase gene of *Clostridium symbiosum*: cloning by polymerase chain reaction, sequence analysis and over-expression in *Escherichia coli*, *European Journal of Biochemistry*, 206, 151.

K. L. Britton, P. J. Baker, P. C. Engel, D. W. Rice, and T. J. Stillman (1993), Evolution of substrate diversity in the superfamily of amino acid dehydrogenases: prospects for rational chiral synthesis, *Journal of Molecular Biology*, 234, 938.

X.-G. Wang, K. L. Britton, T. J. Stillman, D. W. Rice, and P. C. Engel (2001), Conversion of a glutamate dehydrogenase into methionine/ norleucine dehydrogenase by site-directed mutagenesis, *European Journal of Biochemistry*, 268, 5791.

S. Y. K. Seah, K. L. Britton, D. W. Rice, Y. Asano, and P. C. Engel (2003), Kinetic analysis of phenylalanine dehydrogenase mutants designed for aliphatic amino acid dehydrogenase activity with guidance from homology-based modelling, *European Journal of Biochemistry*, 270, 4628.

P. Busca, F. Paradisi, E. Moynihan, A. R. Maguire, and P. C. Engel (2004), Enantioselective synthesis of non-natural amino acids using phenylalanine dehydrogenases modified by site-directed mutagenesis, *Organic and Biomolecular Chemistry*, 2, 2684.

M. J. Abrahamson, E. Vázquez-Figueroa, N. B. Woodall, J. C. Moore, and A. S. Bommarius (2012), Development of an amine dehydrogenase for synthesis of chiral amines, *Angewandte Chemie* (International Edition), 51, 3969.

K. B. Mullis (1990), The unusual origin of the polymerase chain reaction, *Scientific American*, 262, 56.

W. Arber and S. Linn (1969), DNA modification and restriction, *Annual Review of Biochemistry*, 38, 467.

R. J. Roberts (1976), Restriction endonucleases, *CRC Critical Reviews in Biochemistry*, 4, 123.

F. Zhang, Y. Wen, and X. Guo (2014), CRISPR/Cas9 for genome editing: progress, implications and challenges, *Human Molecular Genetics*, 23 (R1), R40.

Further reading

Enzymes (1930) by J. B. S. Haldane. Reissued in paperback by MIT Press, 1965. A classic reflecting the state of thinking about enzymes ahead of any knowledge about their chemistry and structure.

Introduction to Protein Structure 2nd Edition (1999) by Carl Brändén and John Tooze, Garland Publishing. A readable, well-illustrated account.

Annual Reviews of Biochemistry, 55, 1–28 (1988) Sequences, sequences and sequences. This review journal always starts with a biographical account of an eminent scientist's contribution. This is Frederick Sanger's account of the detective trail to obtain the first protein sequence.

Max Perutz and the Secret of Life (2010) by Georgina Ferry, Random House.

Outline of Crystallography for Biologists (2002) by David Blow, Oxford University Press. A gentle introduction to a tough subject.

Linus Pauling: And the Chemistry of Life (2000) by Tom Hager, Oxford University Press.

Catalysis in Chemistry and Enzymology (1987) by William P. Jencks, Dover Publications. An excellent, clear account of catalysis.

An Introduction to Enzyme and Coenzyme Chemistry (1997) by Tim Bugg, Blackwell Science. Attractively presented from a chemist's standpoint.

Structure and Mechanism in Protein Science (1999) by Alan Fersht, W. H. Freeman. A more advanced coverage.

Fundamentals of Enzymology, 3rd Edition (2009) by Nicholas Price and Lewis Stevens, Oxford University Press. Enzymology from the biochemist's point of view.

Index

For the benefit of digital users, indexed terms that span two pages (e.g., 52–53) may, on occasion, appear on only one of those pages.

Enzymes

W

X

Y

Z

SCIENTIFIC REVOLUTION
A Very Short Introduction
Lawrence M. Principe

In this *Very Short Introduction* Lawrence M. Principe explores the exciting developments in the sciences of the stars (astronomy, astrology, and cosmology), the sciences of earth (geography, geology, hydraulics, pneumatics), the sciences of matter and motion (alchemy, chemistry, kinematics, physics), the sciences of life (medicine, anatomy, biology, zoology), and much more. The story is told from the perspective of the historical characters themselves, emphasizing their background, context, reasoning, and motivations, and dispelling well-worn myths about the history of science.

www.oup.com/vsi

RELATIVITY
A Very Short Introduction
Russell Stannard

100 years ago, Einstein's theory of relativity shattered the world of physics. Our comforting Newtonian ideas of space and time were replaced by bizarre and counterintuitive conclusions: if you move at high speed, time slows down, space squashes up and you get heavier; travel fast enough and you could weigh as much as a jumbo jet, be squashed thinner than a CD without feeling a thing - and live for ever. And that was just the Special Theory. With the General Theory came even stranger ideas of curved space-time, and changed our understanding of gravity and the cosmos. This authoritative and entertaining *Very Short Introduction* makes the theory of relativity accessible and understandable. Using very little mathematics, Russell Stannard explains the important concepts of relativity, from E=mc2 to black holes, and explores the theory's impact on science and on our understanding of the universe.

RACISM
A Very Short Introduction
Ali Rattansi

From subtle discrimination in everyday life and scandals in politics, to incidents like lynchings in the American South, cultural imperialism, and 'ethnic cleansing', racism exists in many different forms, in almost every facet of society. But what actually is race? How has racism come to be so firmly established? Why do so few people actually admit to being racist? How are race, ethnicity, and xenophobia related? This book reincorporates the latest research to demystify the subject of racism and explore its history, science, and culture. It sheds light not only on how racism has evolved since its earliest beginnings, but will also explore the numerous embodiments of racism, highlighting the paradox of its survival, despite the scientific discrediting of the notion of 'race' with the latest advances in genetics.

www.oup.com/vsi

PLANETS
A Very Short Introduction
David A. Rothery

This *Very Short Introduction* looks deep into space and describes the worlds that make up our Solar System: terrestrial planets, giant planets, dwarf planets and various other objects such as satellites (moons), asteroids and Trans-Neptunian objects. It considers how our knowledge has advanced over the centuries, and how it has expanded at a growing rate in recent years. David A. Rothery gives an overview of the origin, nature, and evolution of our Solar System, including the controversial issues of what qualifies as a planet, and what conditions are required for a planetary body to be habitable by life. He looks at rocky planets and the Moon, giant planets and their satellites, and how the surfaces have been sculpted by geology, weather, and impacts.

"The writing style is exceptionally clear and pricise"

Astronomy Now

THE U.S CONGRESS
A Very Short Introduction
Donald Richie

The world's most powerful national legislature, the U. S. Congress, remains hazy as an institution. This *Very Short Introduction* to Congress highlights the rules, precedents, and practices of the Senate and House of Representatives, and offers glimpses into their committees and floor proceedings to reveal the complex processes by which they enact legislation. In *The U.S. Congress*, Donald A. Ritchie, a congressional historian for more than thirty years, takes readers on a fascinating, behind-the-scenes tour of Capitol Hill-pointing out the key players, explaining their behaviour, and translating parliamentary language into plain English.

www.oup.com/vsi

THE EUROPEAN UNION
A Very Short Introduction
John Pinder & Simon Usherwood

This *Very Short Introduction* explains the European Union in plain English. Fully updated for 2007 to include controversial and current topics such as the Euro currency, the EU's enlargement, and its role in ongoing world affairs, this accessible guide shows how and why the EU has developed from 1950 to the present. Covering a range of topics from the Union's early history and the ongoing interplay between 'eurosceptics' and federalists, to the single market, agriculture, and the environment, the authors examine the successes and failures of the EU, and explain the choices that lie ahead in the 21st century.

www.oup.com/vsi

THE AMERICAN PRESIDENCY
A Very Short Introduction
Charles O. Jones

This marvellously concise survey is packed with information about the presidency, some of it quite surprising. We learn, for example, that the Founders adopted the word "president" over "governor" and other alternatives because it suggested a light hand, as in one who presides, rather than rules. Indeed, the Constitutional Convention first agreed to a weak chief executive elected by congress for one seven-year term, later calling for independent election and separation of powers. Jones sheds much light on how assertive leaders, such as Andrew Jackson, Theodore Roosevelt, and FDR enhanced the power of the presidency, and illuminating how such factors as philosophy (Reagan's anti-Communist conservatism), the legacy of previous presidencies (Jimmy Carter following Watergate), relations with Congress, and the impact of outside events have all influenced presidential authority.

> "In this brief but timely book, a leading expert takes us back to the creation of the presidency and insightfully explains the challenges of executive leadership in a separated powers system."
>
> George C. Edwards III, Distinguished Professor of Political Science, Texas A&M University

www.oup.com/vsi